Beef production from silage
and other conserved forages

Longman Handbooks in Agriculture

Series editors:

C. T. Whittemore
K. Simpson

Books published:

C. T. Whittemore: *Lactation of the dairy cow*
C. T. Whittemore: *Pig production – the scientific and practical principles*
A. W. Speedy: *Sheep production – science into practice*
R. H. F. Hunter: *Reproduction of farm animals*
K. Simpson: *Soil*
J. D. Leaver: *Milk production – science and practice*

Beef production from silage

and other conserved forages

J. M. Wilkinson

Agricultural consultant

Longman
London and New York

Longman Group Limited
Longman House, Burnt Mill, Harlow
Essex, England CM20 2JE
Associated companies throughout the world

*Published in the United States of America by Longman
Inc., New York*
© Longman Group Limited 1985

First published 1985

British Library Cataloguing in Publication Data
   Wilkinson, J. M.
Beef production from silage and other conserved
forages. – (Longman handbooks in agriculture)
1. Beef cattle – Feeding and feeds
I. Title 636.2′13   SF203

ISBN 0-582-45581-2

Library of Congress Cataloging in Publication Data
Wilkinson, J. M. (John Michael), 1942 –
   Beef production from silage and other conserved
forages.
   (Longman handbooks in agriculture)
   Bibliography: p.
   Includes index.
   1. Beef cattle – Feeding and feeds.
2. Silage. 3. Hay. I. Title. II. Title:
Conserved forages. III. Series.
SF203.W62 1985   636.2′13
       84-7955

ISBN 0-582-45581-2

Produced by Longman Group (FE) Limited
Printed in Hong Kong

Contents

Preface

'Beef is an experience.' This American advertising slogan captures the essence of both the consumption and the production of beef in the USA. An experience indeed: to see cattle in the feedlots of the mid-West as far as the horizon, and to eat prime steaks, tender and juicy.

But beef is becoming an experience in another sense. Red meat in general, and beef in particular, is increasing in cost relative to other foods. Projections of consumer demand for beef offer producers relatively little hope of future expansion. The consumer now perceives beef as a luxury, rather than as a regular food. For beef to maintain its share of the meat market, costs of production must be rigorously controlled and the quality of the product improved.

This is a book about the production of beef from conserved forages – principally silage and hay. The emphasis is on the achievement of specified performance targets, in defined systems of production, from feeds that have traditionally been considered as being of low quality and only capable of maintaining the weight of the animal. I have attempted to demonstrate that, on the contrary, the potential for obtaining rapid rates of animal growth from conserved forages is surprisingly great.

I am grateful to those who taught me the subject; to my many colleagues and friends who helped by providing information and advice. I thank those who granted permission for me to quote from their work.

I record my thanks to Dr J. H. D. Prescott for his support and encouragement during the preparation of the book.

Finally, my especial thanks go to my wife, Cherol, to whom the book is dedicated. She not only tolerated my absences whilst I wrote, she also typed the script.

J. M. Wilkinson
Marlow, Buckinghamshire April 1983

Chapter 1 Conserved forages as feeds for beef cattle

This book is about the production of beef from silage and other conserved forages. Information on the value of conserved forages as feeds for beef cattle is considered in relation to the requirements by the beef animal for nutrients. Some new ways are outlined by which beef can be produced at relatively low cost to the livestock farmer.

Beef cattle have traditionally played second fiddle to the dairy cow in European livestock production. In north America, by contrast, beef is a more important industry; its producers have political influence, consumers give beef priority when they spend their money on food and the technology of beef production has reached an advanced degree of sophistication, particularly in the large beef feedlots.

The relative importance of beef in relation to agricultural land and the population of breeding cattle is shown in Table 1.1 for western Europe, USA and USSR. In addition, beef output is expressed per head of the human population. Although output of beef per unit area of land is higher in western Europe than in the USA or USSR, this simply reflects the higher population of cows per 100 hectares of utilized agricultural area (UAA) in western Europe. Output per cow is similar in Europe and USA, although somewhat less for the USSR. By contrast, the output of beef per head of the human population is twice as high in the USA as in western Europe or the USSR. Beef is considered a relative luxury in the USSR and in Europe; also, it is relatively vulnerable to economic pressure from other sectors of the livestock industry. In particular, European beef production, by virtue of its interdependence with milk production, is at risk when milk is in oversupply. In the USA, where most of the beef originates

Table 1.1   Production of beef in Western Europe, USA and USSR

	Western Europe (17 countries)	USA	USSR
Beef per 100 ha UAA* (t)	5.07	2.67	1.05
Beef per dairy and beef cow (kg)	198	203	151
Beef per head of human population (kg)	23	56	26

* Utilized agricultural area
Source: from Cunningham, E. P. (1980) In *British Grassland Society Occasional Symposium, No.* 11

in beef breeding herds, the position is quite different. The success or otherwise of beef as a product is more intimately linked to the cost of other meats, and the disposable income of the consumer.

Apart from the cost of the animal at the outset, by far the greatest cost element in beef production is that of feed and, in particular, the amount of high-energy or high-protein concentrate that is used to maintain the weight gain of the animal. Indeed, on a larger scale, the economic fate of beef production world-wide is linked to the availability of cereal grain for use as feed for animals.

One of the main objectives of this book is to show that this link between cereal grain supply on the one hand, and the economic fate of beef production on the other, can be broken with economic advantage to the producer and the consumer. Uncertainty of supply of feed grains on a world scale, coupled with the fact that beef cattle cannot normally compete with other classes of livestock as converters of feed grains into meat, makes a strategy in which beef cattle are given grass and forage crops with low levels of cereal grain particularly attractive.

The ruminant animal is at its most advantageous relative to non-ruminants (pigs and poultry) when it consumes cellulosic feeds, such as grasses. More importantly, it is then complementary, rather than competitive, with the demand for food for the human population. It is complementary because, not only can it live on feed produced on land that is not producing human food, it also can consume the by-products of food production (such as

Table 1.2 Cost of calf, concentrates and forage (% of total) in systems of beef production using calves born in dairy herds.

	System			
	Cereal beef	Maize beef	Grass/ cereal beef	Grass beef
Age at slaughter (months)	11 to 12	15 to 16	16 to 18	22 to 30
Costs				
Calf	30	33	31	29
Concentrates	60	39	44	35
Forage	2	18	16	25

Source: from Allen, D. and Kilkenny, J. B. (1980) *Planned Beef Production*, Granada

cereal straws, sugar beet pulps and vegetable by-products). By exploiting both the principles of beef production and those of forage conservation, beef *can* be produced efficiently, in both temperate and tropical environments, from conserved grasses, forage crops and by-products. The removal of large amounts of cereal grain from the diet of the beef animal need not necessarily result in a reduction in performance to an unacceptably low level; quality beef can be produced from cattle slaughtered at less than 18 months of age in planned systems of production, at a cost that compares very favourably with those incurred in systems that rely heavily on cereal grain as the major source of energy in the diet.

But, despite the link that exists between beef and cereal-grain, it is clear that grass already comprises a high proportion of the diet of beef cattle. In Europe, it has been estimated that almost 90 per cent of the energy requirements of beef cattle are met by grass and grass products. This relatively high proportion arises because grain concentrates tend to be given to cattle in winter, rather than at pasture.

The fact that supplementary concentrates tend to be used with conserved forages and not with grazed pasture illustrates an important feature of beef production; namely, the reduction in the nutritive value of grass and forage crops that often occurs during the conservation process. Losses of dry matter represent losses of nutrients, that is, of plant cell contents that would

Figure 1.1   Beef cattle are at their most advantageous relative to pigs and poultry when they consume grasses

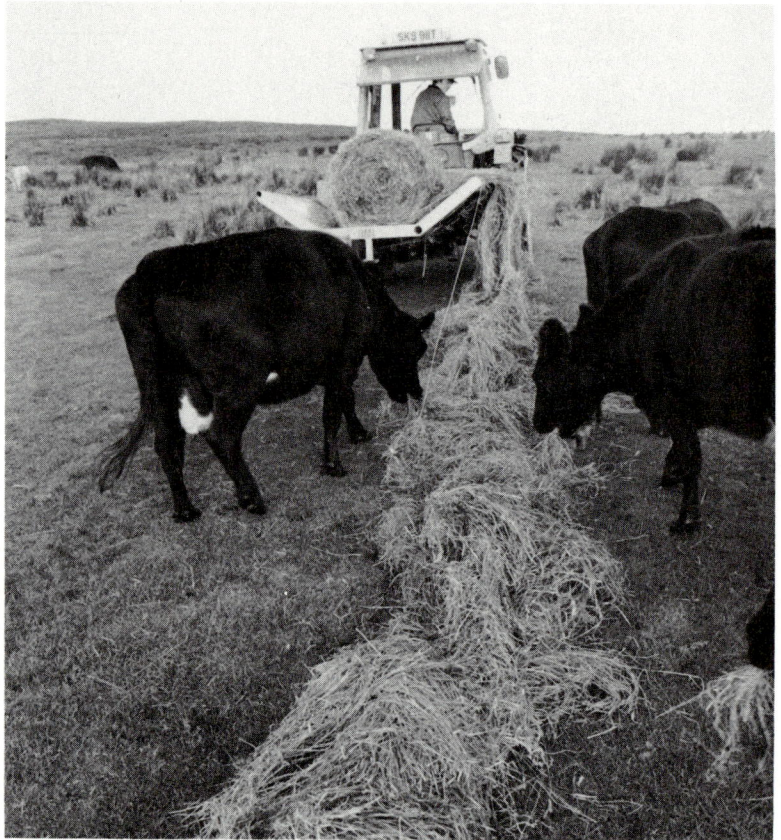

otherwise be completely available for digestion by the animal. The production of high-quality conserved forages for the feeding of the rapidly-growing beef animal is a major challenge to the technical skills of the farmer. One purpose of this book is to show how that challenge can be met. Thus the basic principles of forage conservation are discussed in some detail. They are, after all, fundamental to the success of any beef enterprise in which the intention is to increase, rather than maintain, the weight of the animal during the winter months.

Much progress has already been made in the technology of

forage conservation and in the science of beef cattle nutrition. The days are past when weight gains made in summer were subsequently lost the following winter as cattle scavenged for nutrients amongst foggage, weathered hay, straw, chaff and other by-products that lay around the farm. In consequence, the age and weight at which beef cattle are slaughtered has gradually reduced, and beef output has until recently been sustained by slaughtering an increased number of cattle. If current predictions of a 10–15 per cent decline in the population of dairy cows in Europe are realized and, at the same time, there is a move to a more extreme dairy type, then to maintain beef output it will be necessary to increase weight at slaughter and also to produce more beef from beef breeding, rather than dairy herds. The use of later-maturing breeds of cattle of high growth potential and the choice of feeding systems appropriate to these breeds can lead to

Table 1.3   Components of success in systems of beef production on recorded farms in the UK, 1980

	System*				
	Cereal beef	18-month grass/ cereal beef	24-month grass/ cereal beef	Suckled calf production	Grass- finished store cattle
	percentage contribution to the increase in gross margin per head of top third units compared with the average.				
Lower feed costs	56	70	50	18	−4
Higher live weight at sale	18	9	15	48	48
Lower live weight at start	0	7	0	–	27
Higher price per kg at sale	19	7	16	7	14
Other factors	7	7	19†	27‡	15§

* See Chapter 2 for description of systems
† Lower price per kg at purchase (9%)
‡ More calves weaned per 100 cows (13%), lower cost of herd replacements (11%)
§ Lower price per kg at purchase (11%)

increased weight at slaughter, without necessarily increasing age at slaughter or carcass fatness. Further, there is evidence to show that the adoption of technology that results in good physical performance (low feed costs and high rate of weight gain) are more important determinants of financial success in systems of beef production than buying and selling prices (price per kg at purchase or sale). Thus lower feed costs (principally a lower input of concentrates) and higher live weight at sale accounted for most of the increased gross margin achieved by the top third of recorded beef producers in the UK compared with the average (Table 1.3).

In veiw of the important role that concentrate feeds play in the nutrition of beef cattle given conserved, rather than grazed, forages and also the major effect that concentrates exert on feed costs and on financial success, emphasis is placed on ways in which conserved forages may themselves support growth in beef cattle and so reduce reliance on concentrates.

The book is therefore a statement of a number of technical options open to the beef producer. They are described to illustrate some of the ways in which the beef animal and its supply of feed can interact in contrasting systems of production. No two farms are alike; the techniques and systems outlined here may not be applicable as they stand and will usually require modification to suit individual farm circumstances. They are, however, illustrative of the main principles. The producer may then exploit these to make effective use of his animal and feed resources to produce beef and make money at the same time.

Chapter 2 Systems of beef production

A most important biological principle in considering efficient beef production is that the animal should gain in weight so that it reaches a live weight suitable for slaughter within a reasonable period of time. Thus nutrient intake must exceed requirements for maintenance of body weight, at least for a major part, if not for all, of the animal's life.

The other principle, on which many farmers operate their beef enterprise, is that cattle must be bought and sold 'well'. The difference between making money and simply covering costs can depend to some extent on the skill of the farmer in the market place.

But successful producers not only buy and sell well – they also feed and manage their stock in a systematic, well planned manner in order to be able to market a superior quality product at an above-average price.

Systems of beef production are usually tailored to the individual biological and economic circumstances of the farm. But they fall into broad categories, usually described in terms of the age at which the cattle are slaughtered. Conserved forages feature to a greater or lesser extent in all systems. For example, feedlot beef production comprises the utilization of a high-energy conserved forage – maize silage – together with supplements of maize grain and protein-rich feeds, to produce finished cattle at about 15 months of age. At the other extreme, the over-wintering of beef store cattle often comprises the feeding of low-quality hays or mixtures of hay and arable by-products such as straw. A store animal entering a period of over-wintering such as this at 10–12 months of age is unlikely to reach slaughter weight until it is 2 years of age or more.

Choice of system

Land and feed

The dominant influence on choice of system is availability of feed. Grassland is traditionally used for the production of suckled calves in beef breeding herds. The calves often change ownership at the end of the grazing season because the productivity of the land and its topography prevent the conservation of sufficient feed for both the calf and the cow during the winter months. Commonly, the calves are moved for over-wintering or finishing to arable (grain producing) areas to be housed in feedlots, which can vary in size from a few head to over 100 000.

In milk-producing areas, calves born in dairy herds may be reared at pasture and remain on the farm to receive conserved feeds similar to those given to the dairy herd. Alternatively, they too may be moved to feedlots. Recently, a substantial inter-

Figure 2.1   Weaned suckled calves are often moved for over-wintering or finishing to feedlots, which can vary in size from a few head to over 100 000 cattle

national trade has developed in Europe in which dairy-bred calves are moved at an early age to the grain-producing areas of Italy and Germany for feedlot finishing.

Having chosen a system of production that suits the particular charactaristics of the farm, the producer then has to decide on the appropriate breeds and crosses that suit the system. He also has to determine how many animals to keep, bearing in mind available finance and supplies of feed.

Breeds for feeds

Breeds and crosses of beef cattle show distinctive differences in size, earliness of maturity and carcass characteristics. Large breeds, such as the Charolais, Simmental and South Devon, grow faster than smaller breeds and so produce more beef in a given time. Early-maturing breeds, such as the Aberdeen-Angus, Galloway, Devon and Hereford, finish at a faster rate than late-maturing breeds when given the same feed or they can finish at the same rate on a diet of lower quality. Later-maturing breeds tend to be those of larger body size; they eat more feed than the smaller breeds but are no less efficient at converting feed into body weight gain when slaughtered at the same degree of finish. Breeds differ in their yields of saleable meat, but there is relatively little difference between them in the proportion of high-priced cuts or in the eating quality of the meat.

Examples of the relationship between breed cross and system of production are shown in Fig. 2.2, in which the desired slaughter weight and age of Friesian cattle raised in three contrasting systems are compared with those of Charolais × Friesians (later maturing) and Hereford and Aberdeen-Angus × Friesians (earlier maturing). The earliest-maturing crosses tend not to be used for cereal beef production because they reach an adequate level of carcass fatness at too light a weight. Differences between the earlier- and later-maturing crosses are more pronounced when the cattle are grown on lower-quality diets. Aberdeen-Angus crosses can fatten even on a low-quality store ration.

Although the Hereford × Friesian is capable of growing at a faster rate than the Friesian, in practice the crossbred cattle are given feeds that are of lower quality and are slaughtered at a higher carcass fatness. In this way, beef producers exploit the earlier maturity of the Hereford × Friesian by giving it less-expensive forage diets. By so doing they can produce a higher

Figure 2.2   Relationships between age at slaughter, weight at slaughter and carcass fatness in systems of beef production using calves born to Friesian cows. From MLC (1978) *Breeds for feeds for beef*

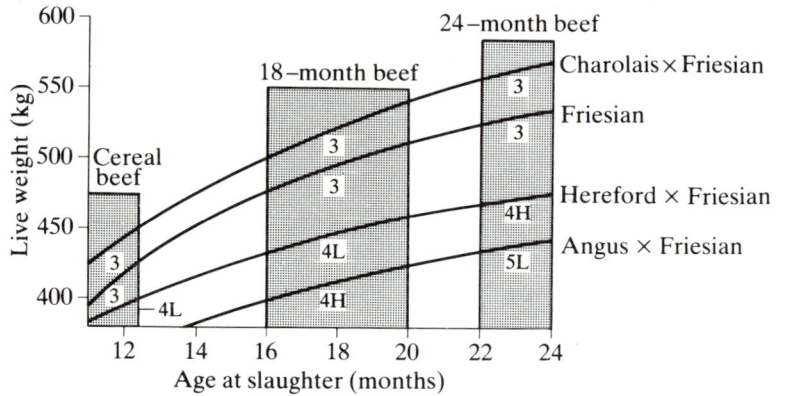

Figure 2.3   Larger breeds and crosses, such as these South Devon × Friesian bulls, are best suited to systems in which high-energy feeds are used

weight of carcass of an acceptable fatness than would be possible if it were given the opportunity of growing to its genetic capacity.

The larger breeds are best suited to systems of production in which high-energy feeds are to be used, such as cereal-grains or maize silage. Because of their high mature body weight, they can be taken to relatively high weights at slaughter without giving carcasses that are over-fat.

Finance

An important factor to be taken into account in the choice of system of production is the requirement for finance to purchase calves, feed, straw bedding, transport and veterinary services. These costs, collectively termed 'working capital', accumulate during the calf's life and reach a peak just prior to sale. For most systems, more than one batch of calves are present on the farm at the same time, thus increasing the peak requirement for working capital (Fig. 2.4). At the outset, the cost of the calf dominates but, as time progresses, feed costs assume greater importance.

The requirement for working capital is least for those systems of production in which the calf is given high-energy, cereal-grain feeds, and slaughtered at a relatively low age and weight. Systems that use calves born in dairy herds and grass-based diets have intermediate requirements for working capital, whilst the production of suckled calves requires almost three times as much working capital per head as cereal beef, largely because of the additional costs of keeping mature females throughout the year (Table 2.1).

The 'gross margin' (sales less variable costs and the cost of the calf) is particularly sensitive to variation in the cost of calves and concentrates; thus cereal beef, developed at a time when both these resources were in plentiful supply, has not generated gross margins as high as those from systems that are based on grass as the major feed and in which the cost of the calf is spread over a higher weight at slaughter (as in the case of 18-month and 24-month beef). Thus, despite lower finance costs, the margin from cereal beef to meet fixed costs and to provide profit is substantially lower than that of 18-month or 24-month beef production. Not surprisingly, these latter systems are more popular amongst farmers.

On the other hand, the relatively lower finance costs associated with cereal or feedlot beef production may be a crucial factor

Figure 2.4   Cash flow profile for 18-month beef production. The peak working capital is reached when two batches of cattle are on the farm at the same time. Adapted from MLC (1971) *Handbook, No. 1*

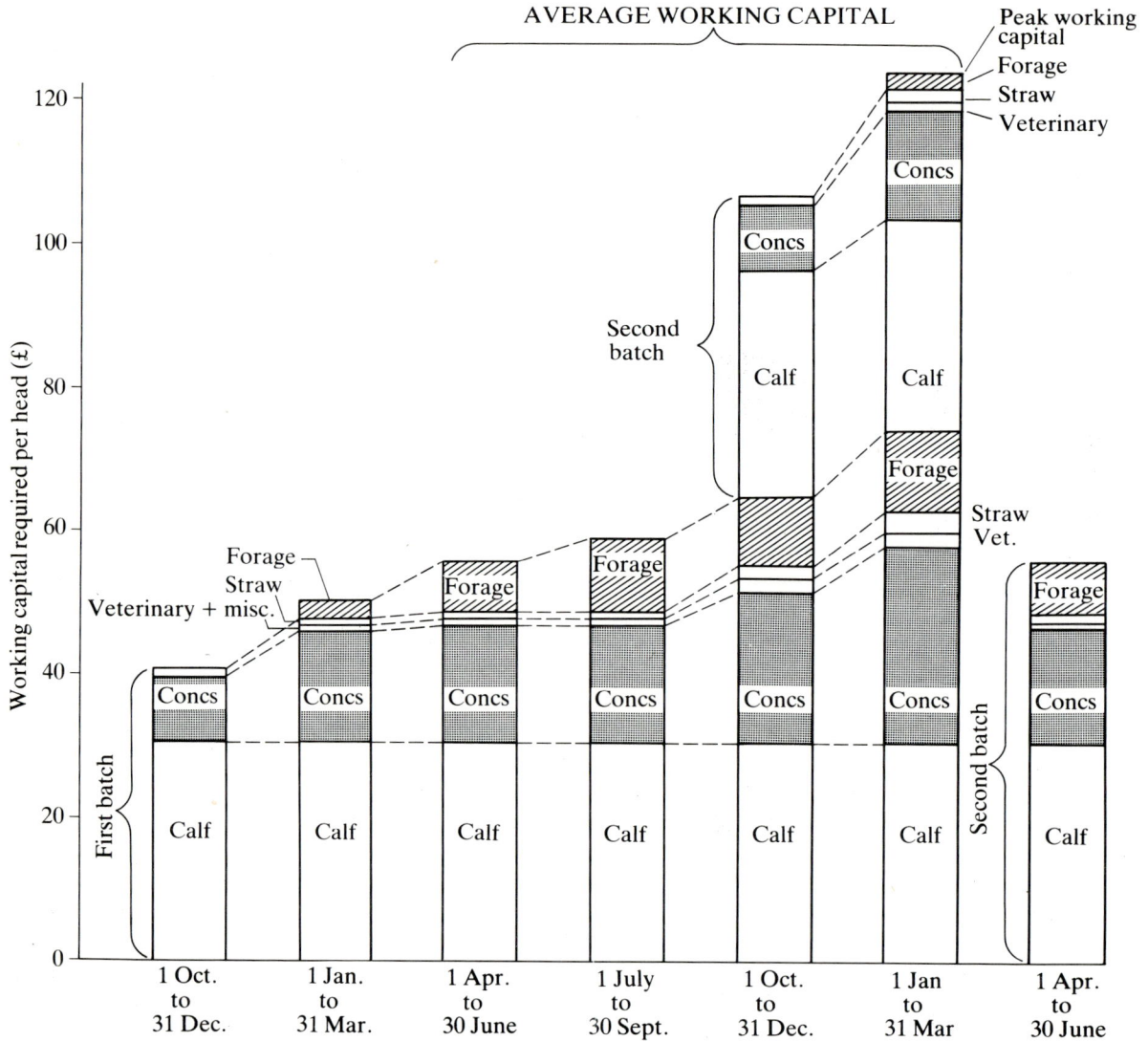

Table 2.1 Average requirements per head for working capital, gross margin, finance costs and margin to meet fixed costs and provide profit from different systems of beef production (cereal beef = 100; based on recorded units in the UK, 1972 to 1977)

	System			
	Cereal beef	18-month grass/ cereal beef	24-month grass beef	Suckled calf production
Average working capital per year	100	138	166	287
Gross margin	100	264	229	243
Gross margin as a percentage of working capital	100	190	137	83
Finance costs*	100	144	169	288
Margin to meet fixed costs and provide profit	100	425	308	183

* Annual payments on loan (interest and capital) to make repayment over a 10-year period

when interest charges are high. There is the further advantage of not having a seasonal pattern of production associated with grazing. Thus the purchase of calves and the sale of finished cattle may be spread over the whole year, with consequent reductions in the peak requirement for working capital, once the cycle of purchase and sale is fully established.

Beef systems   Despite the fact that beef cattle may change hands once or more in their lifetime, it is useful to consider the production of beef as occurring in definable systems, so that targets for production and management can be set, and performance can be monitored. Careful attention to these aspects of the business contributes more to financial success that simply the price at which the animals are bought or sold.

The contrasting systems of production are outlined in Table 2.2, together with details of the usual breed type, major feeds and age at slaughter.

Table 2.2   Systems of beef production

System	Breed	Major feeds	Age of slaughter
Suckled calf production	Beef cross	Grass	9–11 months (weaning)
Cereal beef	Friesian Large beef breed × Friesian	Grain	11–12 months
Feedlot beef	Beef cross (suckled calf)	Grain Conserved forage Grass	14–16 months
Silage beef	Friesian Beef × Friesian	Grass silage Maize silage	14–17 months
Arable beef	Beef cross (suckled calf) Beef × Friesian	Conserved forage Conserved by-products	16–24 months
Grass/cereal beef	Friesian Beef × Friesian	Grazed grass Conserved forage Grain	15–18 months
Grass beef	Friesian Beef × Friesian Beef cross (suckled calf)	Grazed grass Conserved forage	20–24 months (dairy calf) 16–24 months (beef calf)

Clearly, there are many variations of these systems in practice, but a pattern may be detected with respect to the three major characteristics of a system. Thus an earlier age at slaughter reflects higher-energy feeds (cereal-grain). Later-maturing cattle from large breeds are also usually finished on such diets because they do so without laying down excessive amounts of fat in the carcass. Crossbred beef-type calves, often produced in suckler herds, may receive a range of feeds after weaning. Typically in north America, they are transferred to feedlots where they receive grain-based diets, supplemented by conserved forages. In Europe these animals may pass through a 'store' period of relatively slow growth, followed by a finishing period on conserved forages with cereal supplements, or arable by-products. Recently, systems based on maize silage and grass silage have developed, particularly to suit dairy-bred cattle that are slaughtered at an age intermediate between cereal beef and grass beef systems.

Performance targets   The key targets for performance in beef systems are live-weight
gain (which influences efficiency of feed use, total feed consumption and age at slaughter) and weight at slaughter (which influences efficiency of feed use, carcass composition and the value
of the output from the system).

Performance targets are outlined in Table 2.3 for the systems
described in Table 2.2. In both tables the values refer to breeds
and crosses of medium size (except in the case of cereal beef,
where large beef breeds tend to predominate). In the cases of
cereal beef, feedlot beef (the finishing of weaned suckled calves)
and for silage beef, they also refer to bulls.

The successful production of suckled calves to weaning involves
correct management of the cow in relation to season of calving;
the cow's reproductive efficiency largely determines the number

Table 2.3 Peformance targets for different systems of beef production

| System | Performance targets | | | | |
| | Live-weight gain (kg/day) | Weight at slaughter (kg) | Land area (ha per head) | Conserved forage | Concentrates |
				(t dry matter per head)	
Suckled calf production	0.9	250–300[*][†]	0.60 (grass)[†] 0.06 (concs)	1.3	0.2
Cereal beef[†]	1.2	450	0.02 (grass) 0.51 (concs)	0.2	1.8
Feedlot beef[‡]	1.0	460	0.08 (grass) 0.13 (concs)	0.7	0.5
Silage beef[‡]	1.0	500	0.20 (silage) 0.13 (concs)	1.8	0.5
Arable beef	0.8	440	0.13 (forage) 0.05 (concs)	1.2[;§]	0.2
Grass/cereal beef	0.8	475	0.30 (grass) 0.20 (concs)	1.1	0.7
Grass beef	0.7	495	0.40 (grass) 0.15 (concs)	0.7	0.6

* Weight at weaning, autumn calving herds: lower weight, Aberdeen-Angus sire; higher weight, Charolais sire
† Upland herds
‡ Bulls
§ Or equivalent metabolizable energy (ME) as arable by-products or roots.

of calves born and the cow produces most of the feed needed to produce the weaned calf.

Compact calving should be the aim (i.e. calves are born within a relatively short time-period) because, not only it reflect a high conception rate, it also enables the herd to be managed more easily.

In their *Blueprints for Beef*, the Meat and Livestock Commission propose that the feeding of the suckler cow should be aimed at achieving target levels of body condition at key stages in the production cycle, especially at mating (see Table 2.4 and Fig. 2.5).

Table 2.4   *Target body condition scores, on a scale 1 (very thin) to 5 (very fat)*

	Mating	Mid-pregnancy	Calving
Autumn – calving	2.5	2	3
Spring – calving	2–2.5	3	2.5

With cereal beef, feedlot beef and silage beef the cattle are housed throughout the feeding period. Achievement of the performance targets involves control of diseases, especially respiratory diseases, by having correct ventilation of buildings. Bloat may also be a problem in the case of cereal and feedlot beef. Its incidence can be minimized by giving the animals 1 kg of hay or straw daily.

The grass/cereal beef system (also known as 18-month beef) is particularly well adapted to calves born in the autumn. They are housed until 6 months of age, then grazed on land that is also cut for silage. Grassland management is designed to integrate these two activities so that about two-thirds of the total grass area is set aside for first-cut silage. In mid-season the cattle are wormed and transferred to 'clean' regrowth herbage, and the initial grazed area is cut for silage. Supplementary feeds are introduced in late season to maintain growth rate. The level of supplementation depends on pasture availability and its quality. On housing, at at around 12 months of age, the animals are given conserved forage with cereal concentrate supplements. The quantity of concentrates depends on the amount and quality of the conserved forage.

Friesian-type cattle are most commonly used for 18-month beef production. Hereford × Friesians are earlier maturing than pure

Figure 2.5   Examples of body condition scores 1, 2, 3 and 4 (left to right). Courtesy of the East of Scotland College of Agriculture

Friesians and have lower requirements for feed in the second winter period; the calves are, however, more expensive than Friesians.

Grass beef is an alternative to grass/cereal beef, which is more suited to calves born in mid-winter or spring. Animals are usually slaughtered at 20–24 months of age. The emphasis is on maximum use of grass although, because the cattle are older than those reared in the 18-month system, the total requirement for concentrates is similar (Table 2.3). Calves, turned out to pasture when grass is available, may be given supplementary feeds throughout their first grazing season to maintain a target rate of live-weight gain of 0.7 kg/day during this relatively early stage of life. The cattle are housed in the autumn and fed to support a gain of 0.5 kg/day. At turnout to grass for the second grazing season, the aim is to have a rapid finishing growth rate, exploiting compensatory growth, with cattle being slaughtered from July

Figure 2.6 Friesian steers being finished on the grass/cereal system of production

onwards as grass availability declines. Hereford × Friesian cattle are better suited to the system then pure Friesians or later-maturing crosses.

Arable beef is similar to silage beef and feedlot beef in that the cattle are housed throughout the feeding period. These systems are distinguished by the type of feed that predominates. In the case of arable beef, conserved forages are supplemented by straw, roots or arable by-products, to maintain a target rate of gain of 0.8 kg/day. The type of animal best suited to the system is one that finishes relatively rapidly such as the weaned suckled calf. These animals would be slaughted at 16–24 months of age, depending on quality of diet and breed type.

Meeting performance targets

Targets for physical performance, such as live-weight gain, carcass weight at a particular slaughter age and yield of saleable meat, can be set; but two major constraints determine whether or not the animal will or will not achieve them. First, the feeds must be eaten by the animal. So voluntary intake, particularly in the case of conserved forages, may limit performance to an unacceptably low level. Secondly, the concentration of nutrients in the feeds that are eaten must be sufficient for the product of intake and nutrient concentration to give a supply that meets requirements for the desired level of performance.

The feeding systems adopted in most countries for beef cattle are based either on metabolizable energy (ME) or on net energy (that is, the energy available as a fuel to drive the processes of metabolism, to maintain existing body tissues and to synthesize new tissue). Feeds are generally evaluated in terms of the nutrients that are available for metabolism (i.e. ME or metabolizable protein), although in the USA a common expression of the energy value of feeds is the concentration of total digestible nutrients (TDN), which is the sum of the digestible fibre, protein, ether extract (fat) and other digestible components, such as starch, sugars and organic acids.

The requirement of the animal for nutrients depends on the nature of the beast, its weight, sex and target live-weight gain. It also depends on the nature of the feeds that are on offer, since the efficiency with which ME is used by the animal for weight gain is lower for feeds of lower ME content than for feeds higher in ME. Similarly, the requirement for protein in the diet is greater when the concentration of ME is relatively low, since the extent to which protein is synthesized by the rumen microbial population depends on the supply of ME.

Requirements for energy, protein, minerals and vitamins are discussed in detail in the 1980 publication of the Agricultural Research Council's review *The Nutrient Requirements of Ruminant Livestock*, and it is not the intention to repeat the details here. Rather, examples are given of requirements and these are then set in the context of specific systems of production. Finally, some feed budgets are given in an attempt to illustrate the way in which targets can be met from a knowledge of nutrient requirements and feed composition.

Energy and protein requirements

Examples of AFRC requirements for ME (assuming ME: gross energy (GE) = 0.6), rumen-degradable protein (RDP) and undegraded dietary protein (UDP) for bulls and steers of breeds of medium mature size, and for heifers of large mature size (e.g. Charolais crosses), are in Tables 2.5 and 2.6.

The distinction between sexes of animal and breed type principally reflects differences in the energy content of the weight gain. The total requirements for RDP plus UDP are derived from estimates of the total requirement by the tissues for protein (TP) and together they correspond to the total crude protein (CP) requirement. RDP is derived from ME intake (RDP = 7.8 ME, where ME = MJ/day). UDP is required in the diet if RDP is less than TP. The *degradability* of the CP in the diet (RDP/CP) represents the proportion of feed protein that is likely to be degraded in the rumen (usually to ammonia), and is thus unavailable for digestion in the abomasum and subsequent absorption as amino acids of feed, as opposed to those of microbial origin, in the small intestine. Thus from Table 2.6 the optimum degradability of the CP for a 100 kg bull gaining in weight at

Table 2.5   *Metabolizable energy (ME) requirements (MJ ME per day) of cattle for maintenance and growth*

Bulls: breeds of medium mature size
(e.g. Friesian, Hereford × Friesian)

Live weight (kg)	Live-weight gain (kg/day)		
	0.75	1.00	1.25
100	28	33	39
200	44	50	58
300	57	65	75
400	69	79	90
500	80	92	105
600	91	104	119

Steers: breeds of medium size
Heifers: breeds of large mature size (e.g. Charolais)

Live weight (kg)	Live-weight gain (kg/day)		
	0.75	1.00	1.25
100	28	35	43
200	43	51	62
300	56	67	80
400	68	80	96
500	79	94	112
600	89	106	126

Table 2.6 *Rumen-degradable (RDP) and undegraded dietary (UDP)*
protein requirements (g/day) of cattle for maintenance and growth

Bulls: breeds of medium mature size

Live weight (kg)		Live-weight gain (kg/day)		
		0.75	1.00	1.25
100	RDP	225	260	305
	UDP	135	180	210
200	RDP	340	390	450
	UDP	50	75	90
300	RDP	445	510	585
400	RDP	540	615	705
500	RDP	625	715	820
600	RDP	710	810	925

Steers: breeds of medium mature size
Heifers: breeds of large mature size

Live weight (kg)		Live-weight gain (kg/day)		
		0.75	1.00	1.25
100	RDP	225	270	335
	UDP	115	140	150
200	RDP	335	400	485
	UDP	30	40	30
300	RDP	435	520	620
400	RDP	530	630	750
500	RDP	615	730	870
600	RDP	695	825	980

0.75 kg/day is 225/(225 + 135) = 0.625. For most beef cattle of 300 kg or more, the intake of ME is sufficient to support microbial protein synthesis to meet requirements. In the case of diets of relatively low metabolizability (for example, where the ratio of ME to GE is as low as 0.4, rather than 0.6 as assumed so far), intake of ME will be inadequate to maintain a live-weight gain greater than 0.75 kg/day; thus the requirement for TP is correspondingly lower than at higher rates of growth.

Table 2.7 *Total requirements for ME, CP, RDP, UDP and DM for selected systems of beef production from conserved forages: steers of breeds of medium mature size**

	System			
	Feedlot beef	Grass/ cereal beef	Silage beef	
			Grass	Maize
Performance targets				
Age at slaughter (months)	18	18	15	15
Live-weight gain (kg/day)	0.8[†]	0.8	1.0	1.0
Days on feed	180	540	450	450
Total ME required (MJ) Of which	11 160	27 445	23 805	23 805
Conserved forage	6 445	7 680	14 410	16 710
Grazed grass	—	8 910	—	—
Concentrates	4 715	10 855[‡]	9 395[‡]	7 095[‡]
Total CP required (kg) Of which	86.8	214.3	192.6	192.6
RDP	86.8	186.6	161.3	161.3
UDP	—	27.7[§]	31.3[§]	31.3[§]
Total DM required (kg)				
Conserved forage	644	768	1 440	1 562
Grazed grass	—	775	—	—
Concentrates	410	940[‡]	815[‡]	615[‡]

* It is assumed in calculating requirements that the metabolizability of the diet (ME/GE) is 0.6. The ME concentration of the conserved forage was taken a 10.0 MJ/kg DM for feedlot beef, grass/cereal beef and grass silage beef, and 10.7 MJ/kg DM for maize silage. The concentration of ME in the concentrate supplement was taken as 11.5 MJ/kg DM.
[†] 300 kg live weight at start
[‡] Includes milk substitute
[§] Includes 16.0 kg required during the early calf phase (birth to 100 kg)

The factors affecting the intake of conserved forages are considered in Chapter 4; at this stage it is sufficient to stress that, since the daily requirement of ME to meet specific targets for growth is met by a combination of dry matter (DM) intake and the concentration of ME in the DM, and since conserved forages are usually offered to cattle *ad libitum*, it is vital to be able to assess the likely levels of intake that might be acheived with particular forages, and combinations of forage and supplements.

Estimates of the total requirements for ME and protein are given in Table 2.7 for feedlot beef, silage beef and grass/cereal 18-month beef. The data relates to steers of breeds of medium mature size, such as the Friesian.

Feed budgets

In practice, the concentration of ME in conserved forages varies between years and, within years, between cuts. Also, the yield can vary considerably from year to year. Usually the aim is to make use of most, if not all, of the material in store.

The level of concentrate supplement required to meet performance targets depends on the quality of the conserved forage; it may also have to change according to the quantity available. A relatively small change in daily concentrate allowance can, by altering daily gain, have a big effect on the time taken to reach slaughter weight.

Examples of alternative feeding strategies for the final winter period of the feedlot beef and grass/cereal 18-month systems are in Tables 2.8 and 2.9, respectively, together with the corresponding feed budgets for the period. They are based on the Meat and Livestock Commissions's *Blueprints for Beef.* In the case of feedlot beef, it is important to recognize that animals of large mature size (e.g. Charolais crosses) generally require more concentrates than smaller British crosses. This reflects the longer period required by the larger animal to reach a suitable weight for slaughter. Thus at 0.9 kg live-weight gain per day, large breed crosses are slaughtered some 65 kg heavier and have been kept on feed during the finishing period an extra 25 days. Despite a similar daily ration, the larger animal requires 75 kg more concentrates and 0.5 tonne more silage.

If the supply of silage is limited, a slightly higher level of concentrates can be used to increase the intake of ME and daily gain. This has the effect of shortening the finishing period so that, overall, there is little change in the total requirement for concentrates. By contrast, there is a large reduction in the quantity of silage needed, particularly in the case of the 18-month system, where the feeding period is longer than in the case of feedlot beef (Table 2.9).

Table 2.8   Alternative feeding strategies and feed budgets for feedlot beef (silage contains 9.5 MJ ME per kg DM, given ad libitum)

	British crosses*			Continental[†] (large breed) crosses		
Daily feed						
Rolled barley (kg)	1.5	2.5	3.0	2.5	3.0	3.5
Silage (kg DM)	5.8	5.3	4.8	5.3	4.8	4.3
Performance						
Weight at start (kg)	315	315	315	360	360	360
Daily gain (kg)	0.7	0.8	0.9	0.8	0.9	1.0
Slaughter weight (kg)	490	475	460	530	525	420
Finishing period (days)	210	200	160	210	185	160
Feed budget						
Rolled barley (kg)	315	500	480	525	555	560
Silage (t DM)	1.2	1.0	0.75	1.1	0.9	0.7
(t fresh weight at 25% DM)	4.8	4.2	3.0	4.4	3.5	2.7

* e.g. Hereford × Friesian
† e.g. Charolais × Friesian

The product

The performance targets discussed in the previous sections relate to the production of a 'finished' animal, which has a carcass of adequate fatness and appropriate shape for the particular market in which it is to be traded. Market requirements differ between countries; thus in the UK and USA beef tends to be produced with higher levels of fatness than in, say, the Federal Republic of Germany, Italy or Denmark. Differences between countries do not necessaily reflect customer demand; rather, they indicate the extent to which the producer and the meat trade have responded to the demand by the consumer for lean, tender beef.

In an attempt to describe the composition and shape of beef carcasses in a relatively objective manner, a common carcass classification system has been introduced in the EEC. It recognizes two major features that influence the yield of saleable meat in the carcass: conformation (shape) and fatness. Values in Table 2.10 show these effects for UK cattle.

Table 2.9 Alternative feeding strategies and feed budgets for finishing cattle in the grass/cereal system. (Silage is taken to contain 9.5 MJ ME per DM, and is given ad libitum)

	Breed				
	Friesian			Hereford × Friesian	
Daily feed					
Rolled barley (kg)	2.2	2.6	3.0	2.0	2.4
Silage (kg DM)	5.0	4.5	4.3	5.0	4.8
Performance					
Daily gain (kg)	0.7	0.8	0.9	0.7	0.9
Finishing period (days)	285	220	165	215	155
Slaughter weight (kg)	525	500	475	475	450
Feed budget					
Rolled barley (kg)	627	572	495	430	372
Silage (t DM)	1.4	1.0	0.7	1.1	0.7
(t fresh weight at 25% DM)	5.7	4.0	2.8	4.3	2.9

Carcasses with good conformation usually command a premium over the average unless they are very fat. Conversely, carcasses with poor conformation are discounted, especially if they are also very lean.

Table 2.10 Typical yields of saleable meat (% of carcass weight) for carcasses of different conformation and fatness, classified according to the EEC Beef Carcass Classification Scheme

		Fat class[†]						
		1	2	3	4L	4H	5L	5H
	E				72.5			
	U+				72.0			
Conformation	U				71.7			
class[*]	R	74.5	73.4	72.5	71.0	70.0	69.0	66.5
	0				70.3			
	0−				69.6			
	P				68.9			

* Scale: E (excellent) to P (poor)
† Scale: 1 (very lean) to 5H (very fat)
Source: from *MLC Commerical Beef Production Yearbook* (1981)

The use of a classification scheme for carcasses not only allows quality to be reflected in price; it also enables predictions to be made of the likely level of carcass fatness at a particular live-weight in relation to breed and system of production. Thus in Figure 2.2 it can be seen that Friesian cattle may be expected to reach fat class 3 at 12 months of age if reared on a cereal beef or feedlot system; or at 22–24 months of age if reared on the grass beef system. This fat class will be reached at about 420, 490 or 530 kg live weight for the three systems, respectively.

Breed effects on carcass composition were studied in a comprehensive trial conducted by the UK Meat and Livestock Commission, in which the lifetime performance of crossbred calves was examined in contrasting systems of production. Selected results from MLC trials are in Table 2.11.

In the feedlot beef system, in which suckled calves were finished indoors, Charolais crosses had the highest daily live-weight gain, weight at slaughter and killing-out percentage, whilst the smaller Aberdeen-Angus crosses gained weight with relatively greater efficiency of feed use and gave relatively high yields of saleable meat in the carcass.

With the grass/cereal system, in which the calves were out of

Table 2.11 *Effect of sire breed on live-weight gain, efficiency of feed use and carcass composition in two systems of production: cattle slaughtered at fat class 3 on EEC carcass classification scheme (6 to 8% estimated subcutaneous fat content)*

	System							
	Feedlot beef*				Grass/cereal beef[†]			
Breed of sire[‡]	CH	LM	H	AA	CH	F	H	AA
Live-weight gain								
kg/day	0.80	0.69	0.76	0.74	0.94	0.85	0.85	0.84
kg per 100 kg feed	7.63	7.58	8.33	8.55	10.8	10.8	12.0	11.5
Carcass weight (kg)	268	253	216	203	262	216	196	180
Killing-out percentage	55	55	52	53	52	50	49	49
Saleable meat: bone	4.0	4.2	3.9	4.2	3.7	3.5	3.6	3.8
Saleable meat								
% of carcass weight	73	73	72	73	72	70	71	71

* Winter finishing of suckled calves out of Blue Grey (Shorthorn and Galloway) and Hereford × Friesian cows.
[†] 16-month beef; calves out of Friesian cows.
[‡] Charolais (CH), Limousin (LM), Friesian (F), Hereford (H), Aberdeen-Angus (AA).

Friesian dams, Charolais crosses had a higher overall rate of live-weight gain, whilst Hereford crosses tended to have a higher efficiency of feed use. Slaughter at equal carcass fatness showed the Charolais crosses to have a higher killing-out percentage than the Hereford and Aberdeen-Angus crosses, but not than the purebred Friesians. The ratio of saleable meat to bone was lowest for the purebred Friesians, and amongst the crossbreds was highest for the Aberdeen-Angus. Yield of saleable meat was lower for pure Friesian and for Hereford × Friesian than for the others.

The breed of sire influenced not only carcass weight at similar fatness, but also yield of saleable meat and efficiency.

In part of the trial it was possible to measure feed intake, and Table 2.12 shows results for efficiency of gain in the grass/cereal system and also for the grass beef system. Similar trends were evident in both systems with regard to efficiency of saleable meat gain, with the purebred Friesian cattle giving lower gains than beef cross Friesian cattle. Hereford and Charolais crosses tended to produce saleable meat with a higher efficiency than Aberdeen-Angus crosses.

Table 2.12   Effect of breed of sire on efficiency of live weight and saleable meat gain in two systems of production

	Breed of sire			
	Charolais	Friesian	Hereford	Aberdeen Angus
Grass/cereal beef				
Live-weight gain (g/kg digestible organic matter (DOM) intake)	192	192	217	203
Saleable meat gain (g/kg DOM intake)	72.2	67.3	76.1	71.1
Grass beef				
Live-weight gain (g/kg DOM intake)	160	148	165	151
Saleable meat gain (g/kg DOM intake)	59.7	51.1	57.8	52.9

Source: from Southgate, J. R. *et al.*, (1982) *Animal Production* **34**:155 and 167

Chapter 3 **Forage conservation**

The major objective in conserving grass and forage crops is to harvest and store the materials with the minimum loss of nutrients. But adverse weather can wreak havoc with the best-laid plans, particularly in the production of hay. Thus the value of the conserved product as feed for beef cattle is usually lower than that of the crop at the time it is cut for conservation. This, in turn, is determined by the stage of growth of the crop when it is cut. Reduction of nutrient loss during conservation involves the expenditure of energy, either directly in the form of fuel (as with artificial dehydration or barn-drying of hay) or indirectly in the form of machinery for rapid crop conditioning and harvesting. The machinery currently available for the conservation of grass can also be used for other forage crops and crop by-products. Thus legumes such as lucerne can be harvested and conserved in much the same way as grass. Cereals crops can also be conserved as winter feed. Straw, baled and stored in a manner similar to hay, can be a useful source of winter feed for beef breeding cows and store cattle. However, the predominant source of conserved forage for beef production remains the grass crop.

To appreciate fully the factors affecting efficiency of nutrient preservation and the nutritive value of the conserved product, it is helpful to understand how the grass crop grows and changes in composition as the growing season advances.

Growth of the grass crop

Growth commences with increased daylength and soil temperature. The grass plant grows leaf, then stem, with flower and seed production as the culmination of the process. In physiological terms, reproductive growth is more efficient at capturing solar energy than vegetative (leaf) growth but, from the point of view of the farmer and the beef animal, it is not a major benefit since, not only is this early-season growth so rapid that supply exceeds consumption, it is also associated with decreasing feed value. These changes are shown for the first or primary growth (uninterrupted by grazing or cutting) of S23 perennial ryegrass in Fig. 3.1.

Daily growth rate is at its maximum in mid and late May, as stem elongation and ear emergence (flowering) take place. There is a progressive decline in the digestibility of the organic matter in the dry matter (D-value) as the crop matures. Crude protein content also tends to decline in parallel with D-value. Thus, as yield of dry matter (DM) increases through late May and June, the increase in yield of *digestible* nutrients becomes less. With very mature crops there may even be a decline in yield of digestible organic matter as leaves die and stem digestibility continues to decline.

Different species of grasses and forage crops have different patterns of growth and quality during their primary growth. Some commence growth relatively early in the season (e.g. rye and tall

Figure 3.1   Changes in yield and quality of perennial ryegrass. Although yield of dry matter increases into summer, the digestibility (D)-value decreases such that the yield of digestible organic matter increases at a much slower rate as the season progresses. Particularly marked changes occur at ear emergence

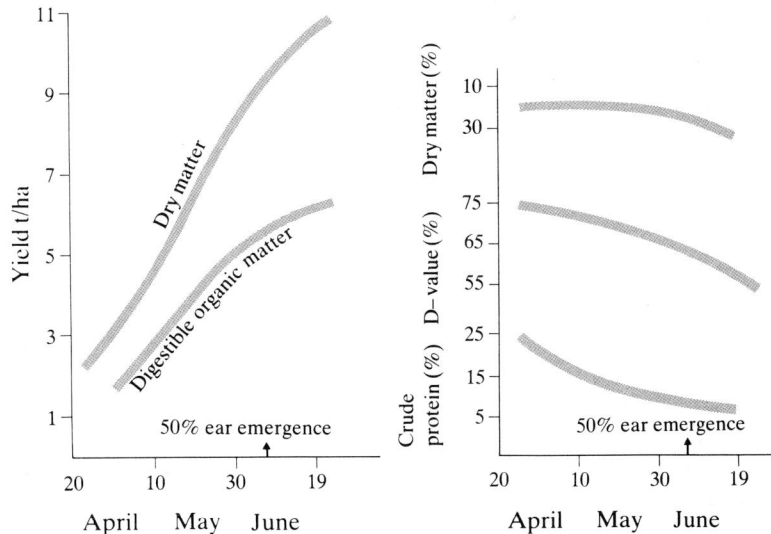

fescue) and show early ear emergence. Others commence growth and flower relatively late in the season (e.g. lucerne). In extreme cases, harvest may not occur until the end of the growing period (e.g. maize). Species and varieties of forage may therefore be chosen to spread the period of time available for harvest at a specific D-value, or may be integrated in a complete harvesting sequence throughout the growing season (see Figs 3.2 and 3.3).

Alternatively, some fields may be grazed in early spring to delay the onset of reproductive growth by 10–14 days, thus increasing the time available for harvest for conservation. This technique is particularly useful in haymaking, when harvest is often deliberately delayed until late June to take advantage of long daylength. Regrowth material may, with early spring grazing or cutting, still contain many stems and have the appearance of a primary growth. The extent to which reproductive growth may be interrupted of prevented depends upon the timing of the early defoliation in relation to the initiation of reproductive growth. If the process has not commenced when the early defoliation occurs, then reproductive growth will proceed in the subsequent regrowth. If early grazing or cutting occurs after the onset of reproductive growth and removes the flower buds, then the subsequent regrowth will be of predominantly leafy material.

Vegetative growth is a particularly useful feature of the grass

Figure 3.2 Grass species and varieties for conservation. By choosing a combination of early-, intermediate- and late-flowering grasses it is possible to harvest the first cut at the same D-value over an extended period. After NIAB (1982) *Technical Leaflet, No. 2*

	MAY					JUNE					SPECIES	EXAMPLE CULTIVARS
	5–10	10–15	15–20	20–25	25–31	1–5	5–10	10–15	15–20	25–30		
		67		63							Cocksfoot (early)	Sylvan
			67		63						Cocksfoot (late)	Cambria
	70		67				63				Perennial ryegrass (early)	Cropper, S24, Reveille
	70		67				63				Timothy (early)	S352
	70			67			63				Italian ryegrass	RVP, Sabalan
		70		67				63			Timothy	S48
		70			67			63			Perennial ryegrass (intermediate)	Talbot, Combi
			70			67			63		Perennial ryegrass (late and very late)	S23, Meltra, Melle, Endura

D–(%) value

Figure 3.3 Yields of dry matter from complementary crops harvested at 67 per cent D-value

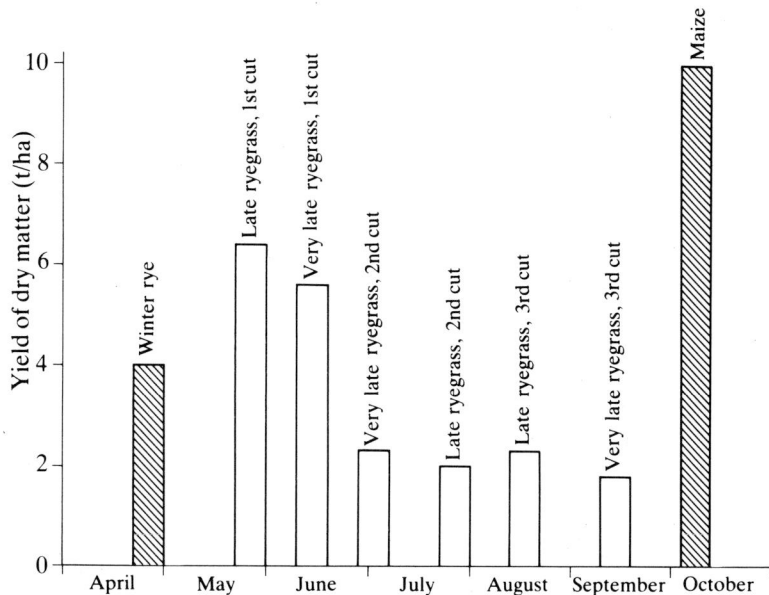

plant because it allows repeated defoliation to occur without removing the growing point of the plant. New growth occurs by the process of tillering. New tillers, which are effectively new plants, are initiated as tiller buds at ground level and can grow to replace the parent plant completely. Legumes do not show this phenomenon, but can protect themselves from death by defoliation by other means. For example, white clover is stoloniferous and, instead of presenting the young growing point for defoliation, the older leaves and petioles tend to be removed because they have grown to a greater height above ground level.

When to cut In Fig. 3.1 the changes are shown in the yield and quality of perennial ryegrass in primary growth. Recommended levels of DM at harvest are given in Chapter 4. The important change is that as yield increases, so quality decreases. Thus the farmer is faced with a dilemma: quantity (high DM yield) or quality (high D-value and crude protein)? A compromise is generally struck between the two, and the extent to which quality is sacrificed for quantity depends on factors such as the method of conservation (silage or hay), the availability of land for conservation in relation to the anticipated requirements of the stock the following winter and the particular beef system in operation on the farm. For example, early and frequent cutting is well suited to artificial dehydration, where there is a requirement to produce a product to a standard specification with respect to digestibility and crude protein for sale off the farm. Frequent cutting for silage can fit in well with a paddock grazing system in which those paddocks that are surplus to requirements are left out of the grazing sequence and cut for silage instead. Late, infrequent cutting is better suited to farms where land is limited or where haymaking is practised. In this latter case, the conserved forage is of relatively low quality and there is no alternative but to supplement the beef animal with cereal grain concentrates in order to produce high rates of weight gain. The farmer is effectively buying in additional land indirectly in the form of supplements.

Late cutting to give a high yield of low-quality conserved forage is well suited to beef systems that do not require rapid weight gains, for example the over-wintering of store cattle. Haymaking

Figure 3.4   A compromise is struck between early cutting, to give high-quality silage with a relatively low yield, and lower-quality silage cut later in the season at a higher yield

is well suited to an infrequent cutting strategy not only because advantage can be taken of long daylength in late June, but also because there is less water in the plant at cutting compared with immature material (Fig. 3.1) Since the price received for hay is rarely related to its nutritive value, yield is economically more important than quality. But here lies a challenge: if nutritional quality could be *improved* during conservation and storage, then high yield could be combined with high quality. Some possibilities for improving feed value of low-quality crops during conservation are discussed in Chapter 8.

Choice of method of conservation

There are important constraints to the choice of conservation system on a particular farm. With high fuel costs there would be little economic sense in investing in an artificial dehydration operation in an area of high rainfall. Equally, the dependence of haymaking on good weather ought also to preclude it from the wetter areas of world. But estimates of hay and silage production in Europe and Scandinavia (Table 3.1) show that farmers commonly make more conserved forage DM as hay than as silage, even in countries with relatively high summer rainfall, such as Ireland and the UK. There are exceptions: in Norway, for example a relatively high rainfall combined with a short growing season has led to a dominance of silage making over haymaking. But the very large amount of hay made in Europe suggests that factors other than climate are involved in choice of conservation system. Capital requirements are greater for silage than for haymaking; thus silage making tends to dominate on larger live-stock farms, where economics of scale can be made, and where large quantities of conserved forage need to be harvested and stored in a limited period of time. The additional capital required to change from haymaking to silage making is illustrated in Table 3.2.

Despite differences in capital requirements in favour of haymaking, silage can be made successfully on smaller farms by either sharing machinery between serveral units or by employing a contractor. In the Netherlands, much of the recent increase in ensilage is attributed to the use of contractors who, by using very large machines, are capable of harvesting and ensiling up to 30 tonnes of crop per hour. This rapid rate of work leads to a better preserved product than might otherwise have been possible using

Table 3.1   Estimated production of hay and silage in western Europe (1978/79)

Country	Hay	Silage
	(Mt dry matter)	
Belgium	1.7	1.9
Denmark	0.4	1.1
Finland	1.7	0.9
France	22.3	16.0
Germany, Federal Republic	9.1	10.5
Ireland	3.6	2.4
Italy	9.3	5.9
Luxembourg	0.2	0.2
Netherlands	1.3	4.1
Norway	0.7	1.1
Sweden	4.6	1.0
United Kingdom	6.9	5.5

Table 3.2   Additional capital required to change from haymaking to silage making on a small livestock farm (1982 prices)

Machinery	£
Extra tractor	9 000
Forage harvester (double chop)	2 775
Extra trailer	1 425
Sides for trailers	750
Push-off back rake	450
Total	14 400

Annual charge
for repairs, depreciation and financing the capital expenditure: £ 5 235

Buildings
Open bunker silo, 500 t capacity: £13 125 gross

the limited resources of machinery and labour at the individual farmer's disposal.

There has been a steady increase in the amount of silage made in recent years, although the increase has not always occurred at the expense of hay. Estimates for the UK, shown in Fig. 3.5, indicate a marked increase since 1970 in the total quantity of DM conserved, with surprisingly little decline in the amount of hay produced.

Surveys have shown that few farmers have completely abandoned haymaking for ensilage. Therefore the most appropriate strategy may be one in which ensiling dominates and in which hay-making is also used, but as a tactical option when the weather is suitable.

Artificial dehydration has remained very much a minor method of forage conservation, not only because of its dependence on fuel oil, but also because of the high cost of capital to establish the dehydration plant. Recent estimates indicate that artificial dehydration accounts for less than 5 per cent of the total DM conserved in the UK. Future developments in engineering technology may mean that dehydration is linked to crop fractionation to produce feedstocks for the chemical industry and residues for livestock feed. Thus the choice of artificial dehydration as a conservation system is less likely to be determined by agricultural

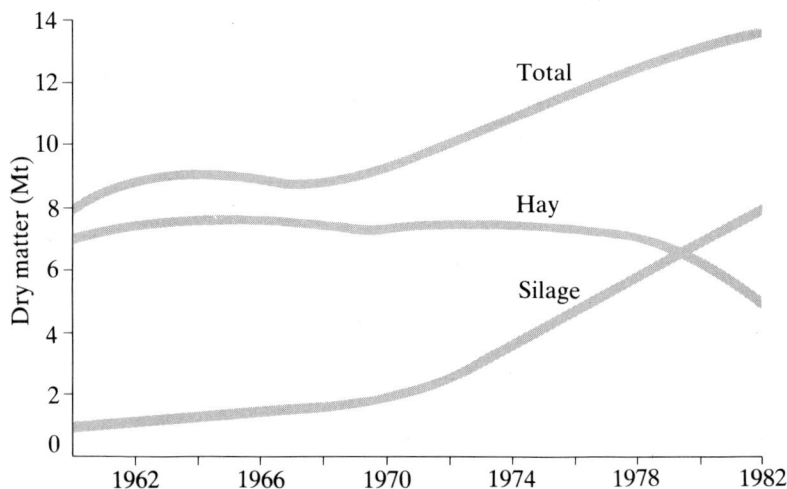

Figure 3.5   Estimated production of conserved grass in the UK

factors than by industrial ones in which grasses and other crops
are seen as an annually renewable source of energy for industry.

In the following three sections, artificial dehydration,
haymaking and ensiling are briefly outlined, with particular
reference to nutrient losses and the factors affecting the feed
value of the conserved products.

Artificial dehydration   The drying of crops artifically rather than by sun-curing involves
the exposure of wet material to a high air temperature, which
rapidly evaporates the moisture. As water is removed, the grass
becomes lighter and the material is drawn further away from the
source of the heat – usually by a fan at the far end of the dryer.
By the time the particles reach the exit of the drying chamber
they are virtually dry, very light and moving rapidly in the air
stream (Fig. 3.6).

Thus the moisture is removed. But before the crop enters the

Figure 3.6   A single-pass high-temperature grass dryer with hammer-mill and
pelleting press. From Raymond, W. F. *et al.* (1978) *Forage
Conservation and Feeding*, Farming Press, Ipswich

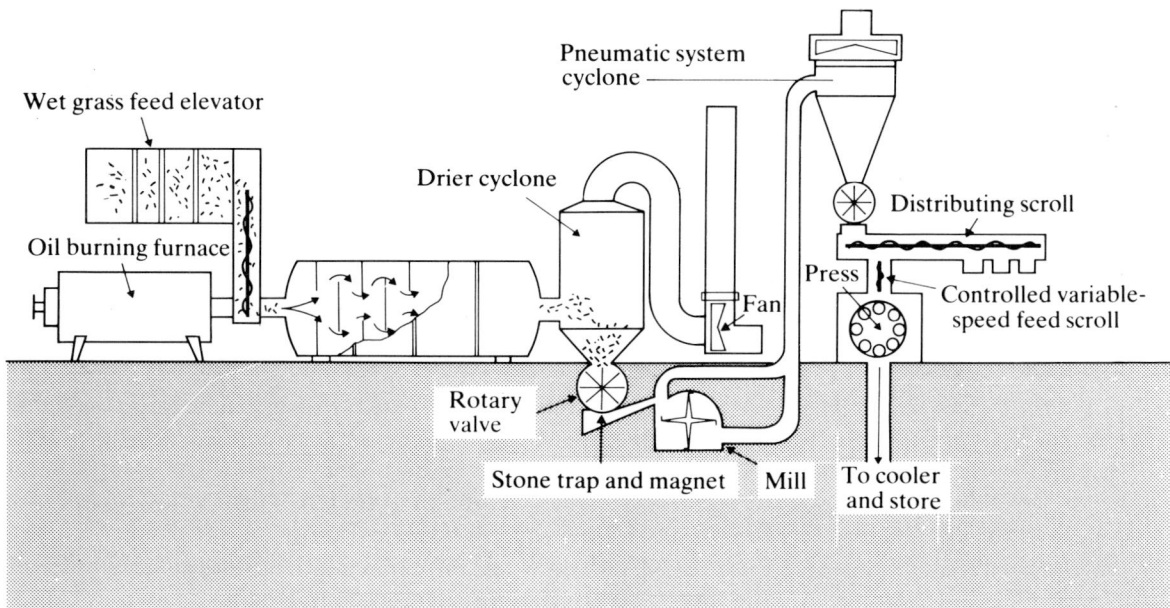

dryer it is usually field-wilted in an attempt to reduce a proportion of the large amount of water in it. Increasing DM content from 18 per cent to 35 per cent by a period of field-wilting reduces the energy required for drying by 60 per cent. Fuel represents the largest single variable cost in artificial dehydration and recent cost increases have highlighted the need to economize in its use. Some possible approaches include the recycling of exhaust gases and burning of straw instead of oil or gas.

A further important treatment of the crop prior to dehydration is short-chopping to a uniform particle length. By reducing the size of particles, the crop can be dried uniformly with reduced risk of catching fire. The chopped material has a greater area of exposed tissue, which enhances moisture loss during drying. A small (7–10 mm) average length of chop facilitates transport of the crop both into, through and out of the dryer, and also allows cubing without having to hammer-mill the material first.

On entry to the dryer the wet crop meets an inlet temperature between 800 and 1 000 °C. In less than 2 seconds it is virtually dry and leaves the drying chamber at 120 to 150 °C with about 15 per cent moisture content. Further processing may comprise grinding by hammer-mill and extrusion through a die to produce pellets, or extrusion alone of unmilled material to produce cobs.

Changes in composition associated with artificial dehydration are relatively small. The most important is the influence of heat on the protein fraction. The denaturation of protein reduces its degradation in the rumen of cattle. Thus the protein in artificially-dried crops, especially the high-protein legumes such as lucerne (alfalfa), is especially useful to growing cattle (see Ch. 5).

Although costly in terms of capital and fuel, artificial dehydration is the most efficient form of forage conservation. DM losses during drying and processing have been estimated at about 2 per cent; to this must be added the losses due to respiration in the field during wilting, at the dryer before drying commences and during subsequent storage. A total loss of DM of 10 per cent probably represents that achieved with good management. As noted earlier, this loss does not appear to relate to the protein fraction; indeed, there is likely to be a significant gain in value in most cases. Thus the major source of nutrient loss is in terms of carbohydrate lost by plant respiration.

Reduced particle size, in pelleted dried forages, coupled with the decrease in ruminal degradation of the protein of dried

Figure 3.7   Effect of pelleting on the net energy of dried forages. Points on the line indicate no difference between chopped and pelleted material; points above the line indicate an improvement associated with pelleting. From Osbourn, D. F. *et al.* (1976) *Proceedings of the Nutrition Society* **35**: 191

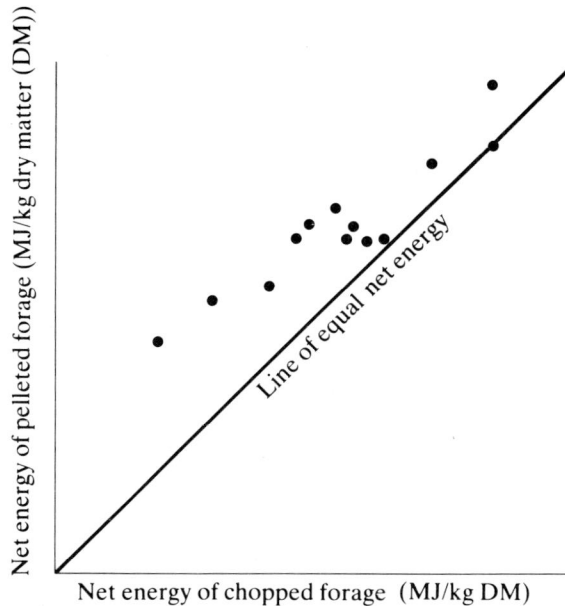

forages, is often reflected in reduced apparent digestibility, but this effect is more than offset by reduced losses of energy during digestion and metabolism. Thus the net energy of pelleted dried forages can be higher than that of the corresponding chopped material (Fig. 3.7). This effect is likely to be greater with crops of lower initial energy value than with crops of high quality. A further effect on feed value of grinding and pelleting the dried crop is that voluntary intake can be markedly increased, particularly with young calves.

Haymaking   The objective of haymaking is to remove water by exposing the wet cut crop to wind and sun, so that it can be baled and stored at low moisture content without further deterioration. In some cases, however, drying may be completed by hot or cold aeration of bales in the barn (barn-drying). The main problem, illustrated

Figure 3.8   Rate of drying in the field of a swath of ryegrass

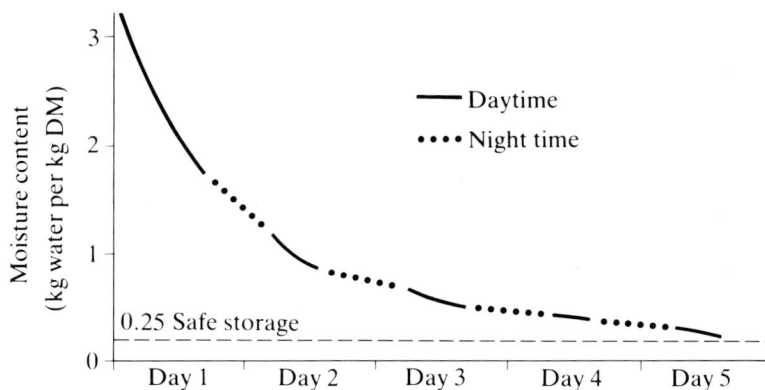

in Fig. 3.8 for a crop of perennial ryegrass drying under field conditions in good weather, is that rate of water loss decreases as drying proceeds.

At cutting there is little resistance on the part of either the leaf or the stem to water loss and drying can continue overnight in dry weather. Plant tissue resistance progressively increases as water has to travel from the innermost parts of the plant to the outside to be evaporated. At the same time, the swath itself becomes unevenly dried, with the surface layer acting as an insulator to prevent the inner parts receiving radiant energy or air to evaporate the water. Turning the swath is therefore most effective in the later stages of drying. Spreading the crop to maximize the surface area and the intercepted radiation is likely to be most effective in the early phase of drying.

Factors affecting drying in the field

Stem-held water delays moisture loss and many farmers now 'condition' the crop at cutting to enhance drying. The effect of crop conditioning on water loss, is particularly marked in the first day after cutting. Grass cut with a mower-conditioner dries faster without further treatment than grass cut with a mower alone and turned once daily.

Ideally, conditioning equipment should scuff and abrade the stem cuticle without damaging the leaf tissue. New developments in the design of mower-conditioners, based on the use of plastic brushes and serrated ribs, should ensure improved drying and reduced field losses.

Other ways of accelerating the loss of water from the plant include the application of dessicants, such as formic acid, and chemicals that are able to draw moisture from the plant by osmosis, such as potassium carbonate. This technique works well with legumes such as alfalfa, where the stem is fully exposed and the leaves lie horizontally, and can therefore be uniformly sprayed with relative ease. Grasses, on the other hand, are difficult to coat evenly with dessicant and it has been found that, although there is an initial advantage in drying rate as the treated leaves dry more rapidly, this is then counterbalanced by slower drying of the stem. It seems that the leaf acts as a wick in drawing water from the stem. If leaf is prematurely dessicated, stem-held water takes even longer to leave the plant. The net result is little overall benefit.

Since leaves lose water at a faster rate than stems, it is hardly surprising that research with crops drying in controlled environments has shown leafy crops to dry at a faster rate than mature, stemmy crops. In practice, however, young crops have a higher initial moisture content (often about 82–85 per cent) than more mature crops (often about 75–80 per cent) (Fig. 3.1) and this, combined with the shorter days of early summer for the drying of young leafy crops, makes it very difficult to achieve more rapid haymaking under field conditions.

A particular problem of ryegrasses is the extent to which the stem is enclosed by leaf sheaths (pseudostems) as the crop reaches ear emergence. Studies have shown quite marked differences in the rate of water loss from crops dried under similar environmental conditions, depending on stage of growth and crop species. The data in Fig. 3.9 show the marked increase in time taken to reach 67 per cent DM content for perennial ryegrass harvested in May as the ears emerged. Tall fescue, on the other hand, dried much faster than either ryegrass or timothy and showed only a small variation in drying time between different harvests. This finding has recently been confirmed under field conditions in the UK.

Clearly, the dominant influence on the rate at which crops dry in the field is the weather. Water is removed by evaporation as a result of radiation, both direct from sunlight and long-wave radiation reflected from clouds. The major effect of radiation is on the surface of the swath, whereas the wind can assist in the process of water evaporation through the mass of herbage in the

Figure 3.9   Effect of species and date of harvest on the time taken to dry the crop to 67 per cent dry matter. From Jones, L. and Prickett, Julia. (1981) *Grass and Forage Science* **36**: 17

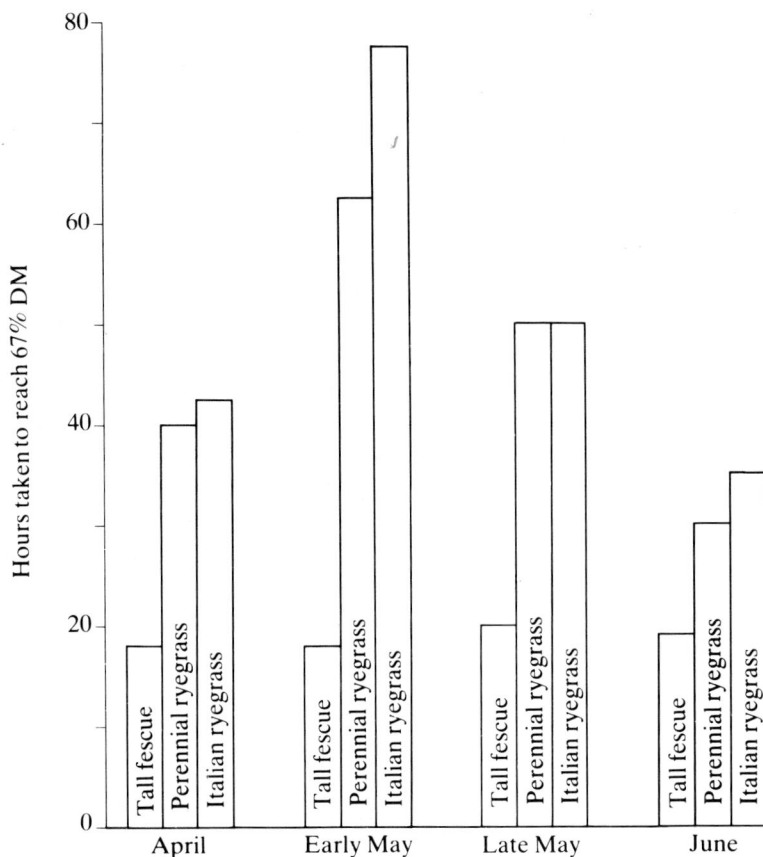

swath. The effect of wind becomes increasingly important as drying proceeds and as the surface of the swath dries out. Wind speed is much greater at the top of a windrowed swath than at ground level; it is clearly valuable to facilitate wind penetration by 'setting-up' the swath.

A useful measure of the drying power of the environment surrounding the swath is the accumulated vapour pressure deficit (VPD). Forecasts of weather and historical records can be used to predict VPD, and in future this may allow a more rational approach to haymaking. Attempts are commonly made to match hay-time with the most likely period of sustained dry weather in summer. Fields are often grazed in early spring before being closed up for hay. Spring grazing may delay the production of stem, but usually does not prevent it; however, digestibility is often higher in crops mown for hay after a spring grazing than in crops of primary growth.

Losses during haymaking

The principal cause of loss of nutrients from the hay crop is plant respiration in the field. Other important sources of loss are mechanical damage during field operations, leaching of plant nutrients as a result of rain, and loss due to continued respiration, bacterial growth and moulding during storage.

Losses due to respiration are difficult to isolate from other field losses, but they are undoubtedly greater in poor weather than when conditions are good for drying. Information from the Federal Republic of Germany in Fig. 3.10 shows not only the dominant influence of weather on total field losses, but also

Figure 3.10   Losses of dry matter during drying in the field

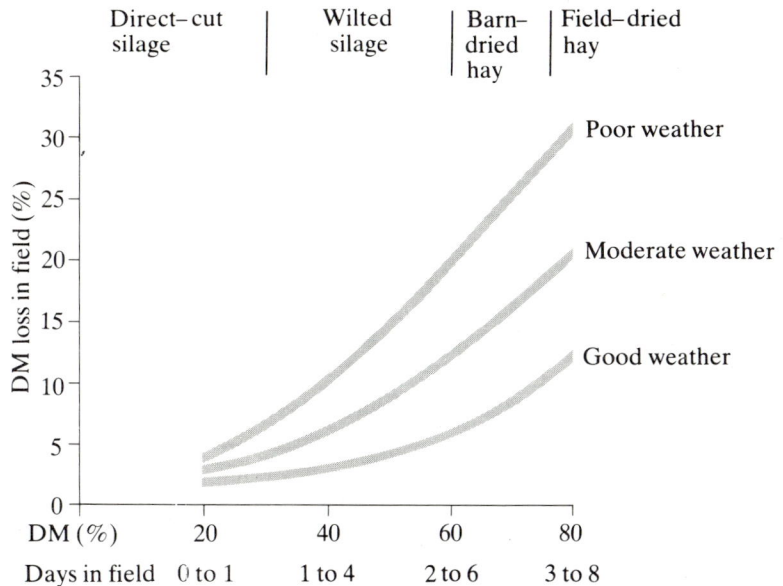

the extent to which losses can increase markedly as the field drying period is extended to produce field-cured hay.

Rain therefore has a large influence on losses. Comparisons between hays made with and without rain show quite marked differences in nutritive value, particularly in field-dried hay made from crops cut at a young stage of growth and of a high initial digestibility.

It is important that the hay crop is baled at low moisture content: 80 per cent DM is the recommended 'safe' level for barn storage. Apart from the risk of heating leading to spontaneous combustion, bacterial and mould growth can include the actinomycetes responsible for farmer's-lung disease. Storage losses with well made hay are likely to be less than 5 per cent of the DM put into the store, unless the crop receives prolonged exposure to the weather during the storage period.

Typical losses during haymaking under conditions of good management are itemized in Table 3.3 for European conditions. The Table shows that losses may be reduced by barn drying or by addition of preservative. Both techniques enable the crop to be removed from the field at a higher moisture content than would otherwise be safe for storage.

Table 3.3   Losses likely to occur from field-dried, barn-dried and ammonia-treated grass hay made under conditions of good management

Dry-matter (DM) loss (%)	Field-dried hay*	Barn-dried hay†	Ammonia-treated hay‡
In field			
Respiration	8	8	8
Mechanical losses	14	4	4
During storage			
Respiration	1	4	–
Fermentation	2	3	3
Surface waste	2	1	–
During removal from store	1	1	1
Total	28	21	16

* 6 days in field, no rain
† Baled at 60% DM
‡ Baled at 60% DM, 35 kg NH_3 per t DM

Barn-drying The technique of completing the drying of hay crops in the barn, developed in the 1950s, involves baling the hay crop from the field at 60–70 per cent DM and then placing the bales either on top of a wire-mesh floor or in the shape of an arch, so that a tunnel is formed underneath. Air is then blown through the bales to complete the drying of the hay.

It is essential that sufficient air (heated or cold) is blown through the complete mass of hay to reduce to a minimum the heat produced by continued plant respiration. The general procedure is to blow air continually through the hay for several days, then to have a period of intermittent drying with careful monitoring of the temperature of the hay. When there is little rise in temperature during periods when no air is being blown, the hay may be removed from the barn dryer and stacked in the same way as field-dried material.

The effect of barn-drying on the nutritive value of hay is principally one of the conservation of a greater content of digestible nutrients in the material compared with swath-made hays. This benefit is much greater when field-dried material receives rain during the final phase of drying than when the weather for field drying is good.

Barn-drying incurs additional fixed costs, which may be under-utilized if the weather is good or if crop yield is low. The fact that barn-drying has remained relatively unpopular is evidence that the extra costs, including the additional labour associated with moving hay into and out of the dryer, are not always recouped in the improved value of the product. There are also problems of matching the capacity of the dryer to the bale-handling system of the farm. Surveys have indicated that most farmers who had decided not to adopt barn-drying had done so because they regarded a dryer as either too expensive or unnecessary. Thus the adoption of tactical methods to reduce losses and improve the quality of hay, which may be used or not depending on actual and forecasted weather conditions, and which do not require large investments of capital or labour, have proved more attractive than barn-drying.

Preservatives for moist hay The addition of chemicals to moist hay to assist in its preservation has been practised intermittently for many years. Salt (sodium chloride) was used at one time to reduce the free-water content of moist hay but the practice has not continued, probably because of the difficulty in applying a powder uniformly to the crop. In addition, there were occasionally problems of digestive upsets in cattle given hay treated with salt.

Recently, attention has focused on the use of propionic acid and its salts as preservatives for hay. Apart from the antimycotic properties of the acid, it has the advantage of being a metabolic product of the fermentation in the rumen and is therefore not toxic or likely to interfere with the processes of digestion.

But there are problems in using propionic acid with moist hay, principally associated with application technology. The acid is volatile and can be metabolized by spoilage organisms if their growth occurs in pockets of hay that have not been adequately treated. Uniformity of application is therefore essential.

Ideally, therefore, the content of water in the crop should be known before a preservative is added, so that the rate of addition can be related to the amount of moisture in the crop (Fig. 3.11)

Propionic acid has been largely superseded by salts such as am-

Figure 3.11   When moist hay is to be treated with preservative prior to baling, the quantity *retained in the bale* should be related to dry-matter content at baling. *Propionic acid equivalent retained in bale.

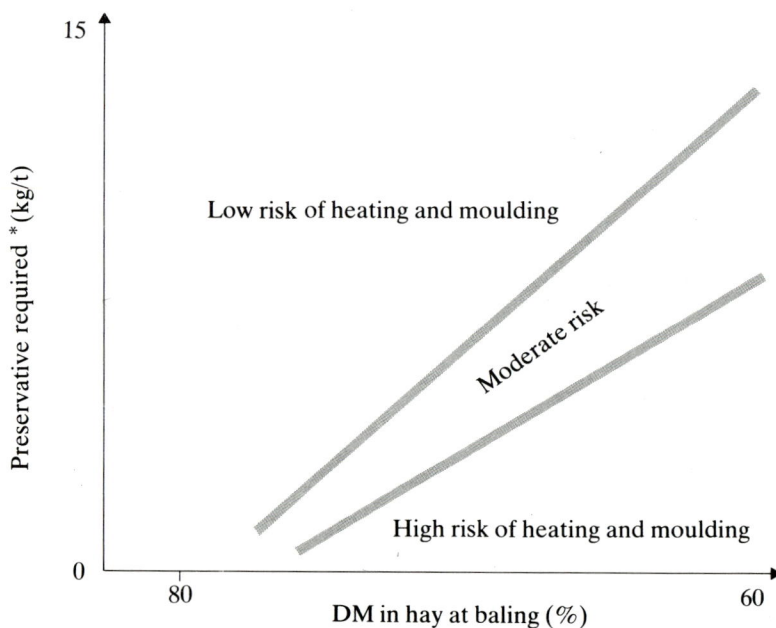

monium propionate and ammonium bispropanoate, which are less volatile. These salts have been found to be almost equally effective as preservatives as the acid itself, suggesting that ammonia also may have antimicrobial properties.

Trials in which moist hay has been effectively preserved, with an adequate quantity of ammonium bispropanoate retained in the bale, have shown that the nutritive value of the stored product is similar to that of barn-dried material (see Ch. 5).

Ensiling

The process of ensilage involves the fermentation of plant sugars to organic acids. The acidity thus produced effectively 'pickles' the crop in a stable state.

Fermentation is an anaerobic process; the crop must be completely sealed from the air to facilitate the growth of the essential anaerobic bacteria that are present in very small numbers on the crop in the field and most probably are introduced to the plant material via the forage harvester. It follows that, as far as possible, air should be removed from the crop at the time of ensiling, by its consolidation in the silo, to exhaust oxygen as rapidly as possible.

Consolidation is normally achieved by chopping or lacerating the crop, either at the time of harvest in the field or at the point of entry to the silo, and by rolling the filled silo with a tractor.

Harvesting

Crops may be cut and removed from the field in one operation by using either a flail harvester, a double-chop harvester or a direct-cut metered-chop harvester. Flail harvesters cost less than double- or metered-chop machines, but they do not chop the crop short.

Often the crop is cut in advance of harvest by a conventional mower. This allows a period of field-wilting to occur which should not exceed 24 hours. There are clear advantages to this procedure, despite the extra work in separately mowing and harvesting the crop. First, there is an opportunity to remove water, which has the effect of increasing the concentration of plant sugars and also reduces the production of effluent from the silo. Secondly, the quantity of DM harvested per day is increased. Thirdly, most metered-chop harvesters and forage wagons are designed to pick up and chop a previously mown swath, in which the plants are aligned parallel to the direction of travel and at right-angles to the cutting blades. This produces a relatively uniform particle

Figure 3.12 Wilted grass being harvested by metered-chop harvester. Note the containers for additive mounted on the chute

Figure 3.13 Harvesting systems for silage. From MAFF (1978) Silage Making

(a) Direct cutting:flail or double–chop harvester

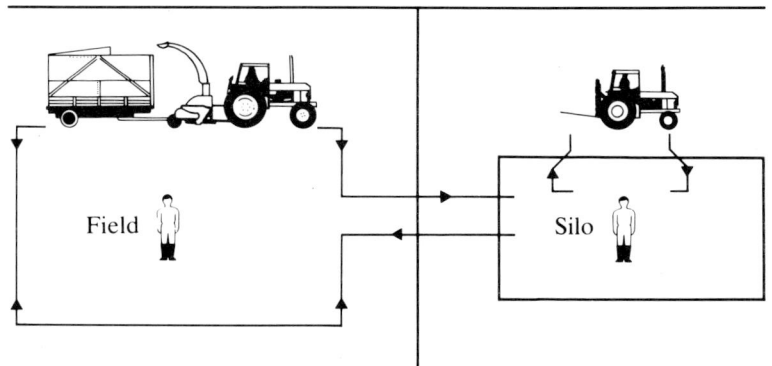

(b) Wilted silage: flail or double–chop harvester

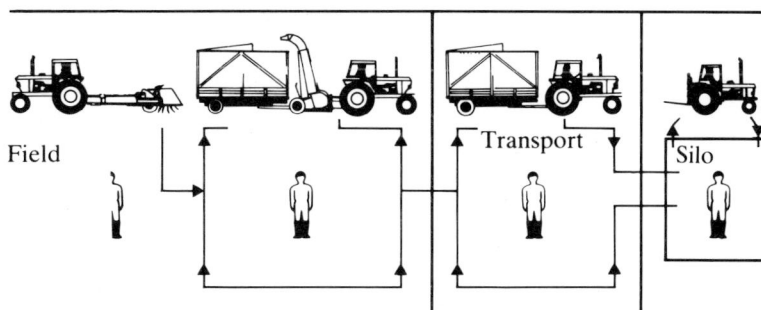

(c) Wilted silage: metered–chop harvester

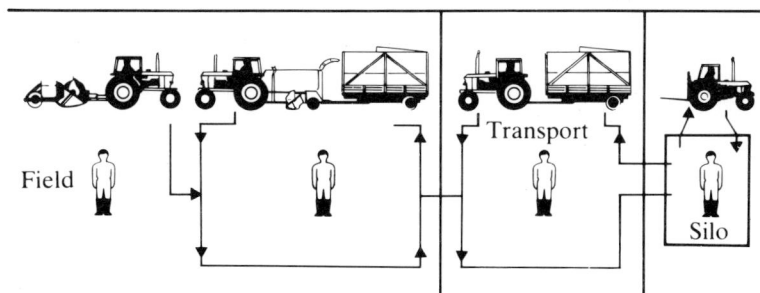

(d) Wilted silage: forage wagon

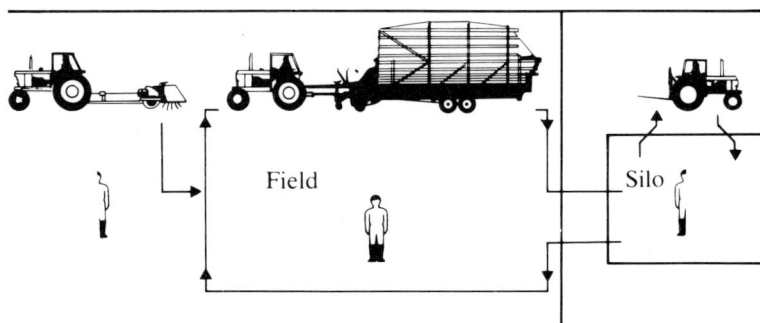

length in the cut crop. With forage wagons, however, chopping of such crops is made more difficult, since the cutting blades are commonly aligned parallel to the direction of travel. Thus some wagons give very little reduction in particle length, although they do have other advantages. They are less costly than forage harvesters, use less energy and are well suited to upland topography. Their popularity is at present confined to smaller farms, which do not have sufficient labour to make 3- or 4-man teams for mowing, harvesting, transport and silo filling.

Some examples of different systems are shown diagrammatically in Fig. 3.13.

Storage

Sealing is effected either by placing a plastic sheet over the crop mass and protecting the sheet from damage (by wind, animals or birds), or by putting the crop into a tower made of steel or concrete.

The most common type of silo is probably the bunker; it is well suited to the storage of large quantities of crop at relatively low cost and therefore tends to be used on large feedlots. Towers are more common on smaller farms, particularly in northern Europe and Scandinavia where greater protection from winter rain and snow is necessary. Silos can be made simply from pits dug in the soil, and these are particularly useful in hilly areas where the silo can be filled from above and emptied from below. Unwalled clamps, covered with plastic, are an attractive alternative to the pit silo, but in each case a concrete floor is recommended to avoid contamination of the silage by soil. Examples of bunker and clamp silos are shown in Fig. 3.14.

Filling should be as rapid as possible, and each day's harvest should be loaded into the silo to form a wedge, with a sloping face, up which the loading tractor runs. At the end of each day's work the exposed surface should be covered with a polyethylene sheet to prevent air movement and heating overnight. As each section of the silo is filled and consolidated by rolling, the side plastic sheet is pulled across the shoulder and the top sheet is sealed to it, using mastic, to form a double seal. It is important that the silo has this double seal over the least well consolidated sides and shoulders, to reduce wastage in these areas. It is equally important that some weighty material (straw bales, car tyres or manure) should be placed over the whole of the exposed plastic sheeting to prevent entry of air, or damage to the sheet by wind, birds or vermin.

Figure 3.14   (a) Silo with pre-cast wall panels (b) Unwalled clamp (c) Earth-walled bunker; such a bunker may also use natural topography and be built into a hillside

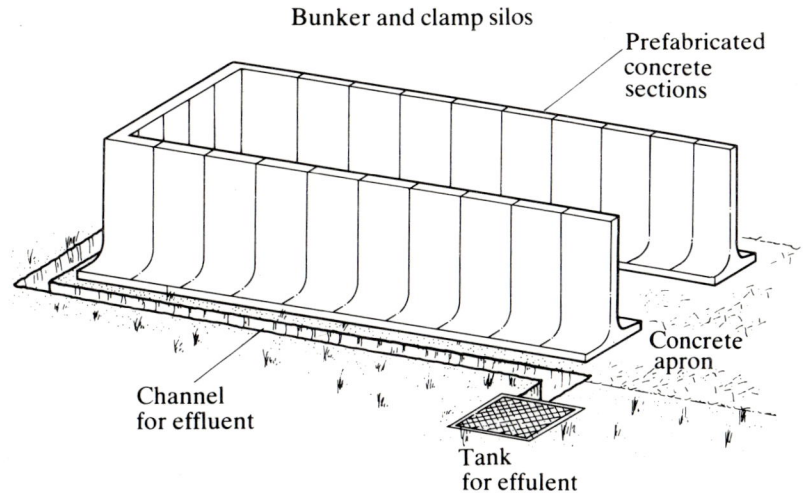

Bunker and clamp silos

Prefabricated concrete sections

Concrete apron

Channel for effluent

Tank for effulent

(a) Unroofed concrete bunker- silo

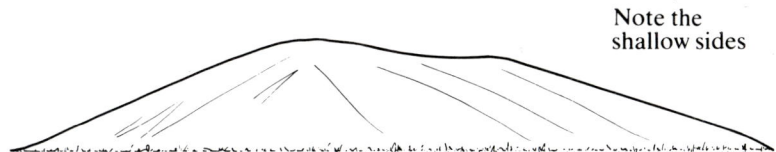

Note the shallow sides

(b) Unwalled clamp

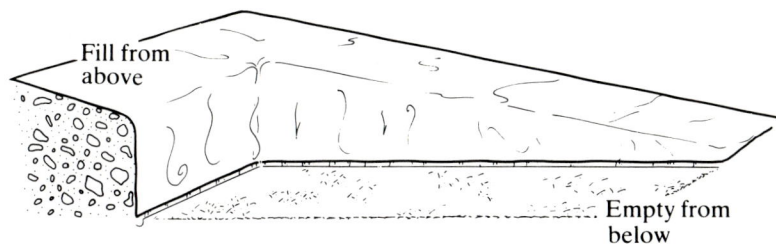

Fill from above

Empty from below

(c) Earth-walled bunker built into a hillside

Figure 3.15   Roofed concrete bunker-silos, with straw bales protecting the polyethylene sheet to ensure complete sealing of the silage

Patterns of fermentation

At the point of entry into the silo, the pH of the crop is usually about 6.0 (slightly below neutrality); the crop is alive, but consuming the products of photosynthesis (sugars) by respiration and producing carbon dioxide. However, as the supply of oxygen to the plant becomes exhausted, fermentation of the sugars to acids dominates. The pH of the crop decreases, since the principal products of fermentation at this stage are lactic acid and acetic acid. During both respiration and fermentation, plant and microbial enzymes degrade the protein fraction to amides and amino acids. Occasionally, this degradation extends to the formation of ammonia during the secondary fermentation of lactic acid to butyric acids by clostridial bacteria.

Both the pattern of fermentation and its extent depend essentially on three main factors: the amount of water in the crop, the amount of fermentable substrate – water-soluble carbohydrates (sugars) – and the capacity of the crop to buffer the reduction in pH brought about by the fermentation.

Water and substrate interact. With relatively dry crops, fermen-

Figure 3.16
Typical values for the pH of silage made from grass crops harvested at different contents of dry matter

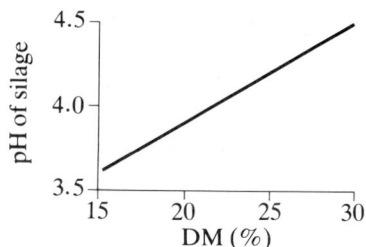

tation occurs to a lesser extent than with wetter crops; bacterial activity ceases as the amount of free water (i.e. that not associated with products of fermentation) is reduced, even though the water-soluble carbohydrates (WSC) may not be completely fermented. The ultimate pH at which the silage remains stable therefore increases with increasing DM content (Fig. 3.16).

On the other hand, wet crops of low WSC content may undergo a secondary fermentation. In this situation there is ample water for bacterial growth but insufficient WSC to generate a vigorous lactic acid-dominated fermentation. The buffering capacity of the crop is usually quite high and these factors combine to give an insufficient decrease in pH to prevent the growth of clostridial bacteria. The growth of these bacteria is generally inhibited below pH 4.5. Their activity, however, results in the fermentation of lactic acid to butyric acid, the complete degradation of proteins and amino acids to ammonia, an increase in the content of acetic acid in the silage, and a gradual increase in pH during the storage period. These changes are shown in Figs 3.17 and 3.18.

The consequences of a secondary, clostridial, fermentation are that the silage is poorly preserved. Compared with well preserved

Figure 3.17 Typical changes in the content of fermentation acids in silage as a result of secondary fermentation

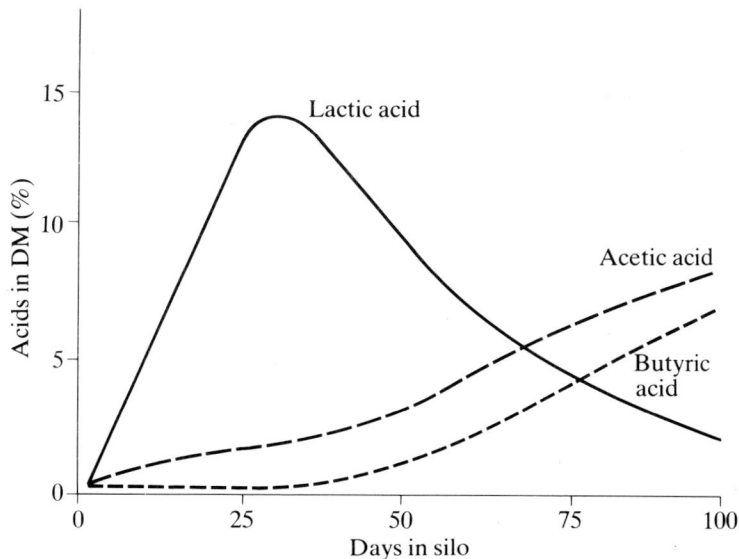

Figure 3.18   Patterns of fermentation and changes in the pH value of silage as a result of secondary fermentation

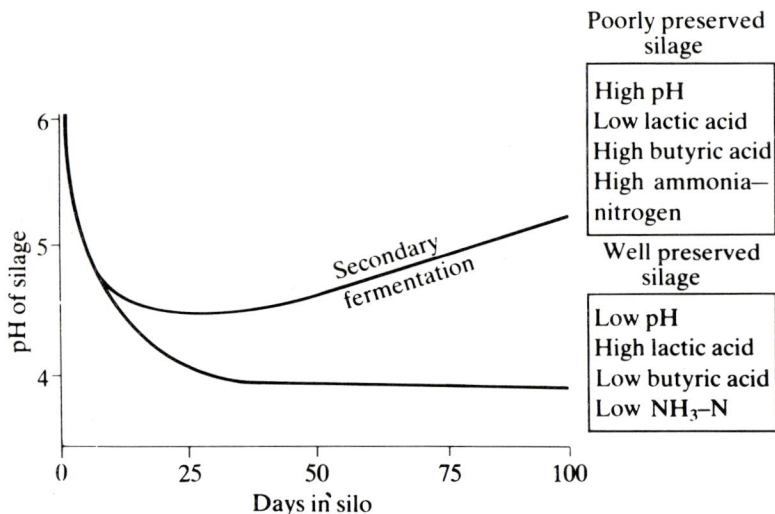

material, such silages show a decrease in energy value and hence a decrease in digestibility, a reduction in the efficiency with which the nitrogenous fraction is utilized by the animal, and a reduction in voluntary intake (see Ch. 6).

The principal factors contributing to a secondary fermentation are low contents of DM and of WSC in the crop at harvest, but a crop of relatively high sugar content, such as maize, is less likely to give rise to a secondary fermentation than one of lower WSC content, even though it may have a low DM content. Grasses tend to have higher contents of WSC than do legumes and temperate grasses generally accumulate higher contents of WSC than do tropical crops.

A useful concept is that of a 'critical' level of WSC, below which an additive is needed to assist in the achievement of a well-preserved silage and above which an additive is not needed. The critical level of WSC is thought to be about 3 per cent of the fresh crop weight. Levels of WSC generally increase as grasses mature, at least until the point of flowering. A guide to the likely levels of WSC in perennial ryegrass in relation to stage of growth and DM content at cutting is given in Fig. 3.19.

It is essential to wilt young grass crops, to concentrate the WSC, particularly if the intention is to ensile without an additive.

Figure 3.19   Likely levels of water-soluble carbohydrates (WSC) in perennial
ryegrass in relation to stage of growth and dry-matter content

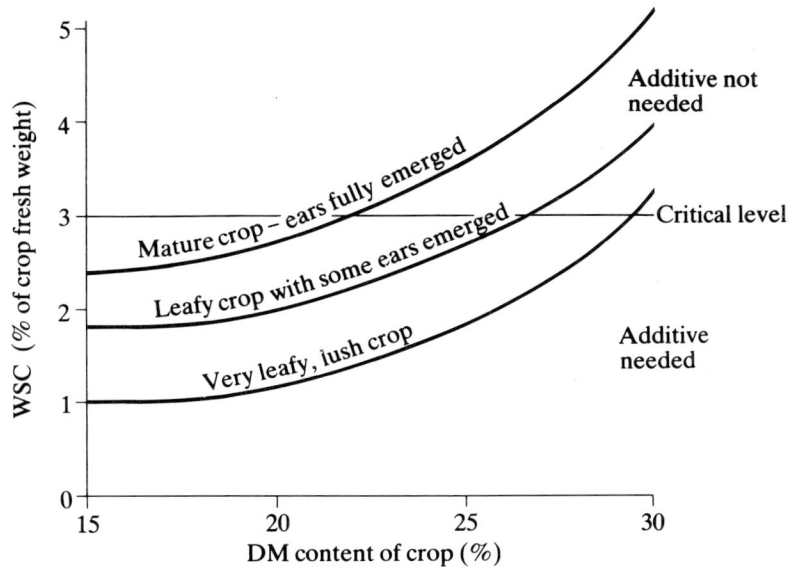

Additives for silage   A particular problem of ensilage is that the quality of the product
cannot be fully assessed until the period of storage is completed
and the silo is opened. Samples can be taken for analysis by
taking cores through the silage mass and re-sealing but, if preservation quality is poor, little can be done to rectify the situation
at the time of feeding other than to use less silage in the ration.

Additives were introduced in an attempt to improve the
predictability of the ensiling process, so that crops that might
otherwise undergo a secondary fermentation would not do so. In
the case of maize, sorghum and tropical crops of low-nitrogen
content, additives such as urea have been used to improve the
nutritive value of the product by rectifying this specific nutrient
deficiency. More recently, additives for grass have been developed that contain formaldehyde. This chemical binds to proteins
to render them less prone to degradation in the rumen. In this
way it is hoped that the supply of amino acids to the animal may
be improved.

But the most common use of additives in silage making is to
improve fermentation quality and thereby reduce nutrient loss.

Table 3.4   Additives for silage: active ingredients and recommended rates of use

	Recommended rate of addition* litres/t	
	Grasses	Legumes
Additives to improve preservation		
Formic acid (85%)	2.5	5.0
Ammonium tetraformate	3.0	6.0
Molasses (50% sugars)	10	15
Additives to improve preservation and reduce protein degradation		
Formalin‡ + formic acid	1 + 2	2 + 3
Formalin + sulphuric acid (50%)	1 + 2	2 + 3
Additives to increase the content of nitrogen		
Urea	10 kg/t	—
Urea + molasses	10 + 10 kg/t	—
Urea + minerals (50%)	20 kg/t	—

* To direct-cut crops
‡ 35% w/w formaldehyde

A list of the active chemicals used in additives and their recommended rates of addition is given in Table 3.4.

Molasses has the advantage of supplying additional substrate to crops low in WSC without the handling and safety problems associated with formalin or acids. It is difficult to apply in cool weather conditions, because of increased viscosity, and the quantity required per tonne to influence significantly the pattern of fermentation is high.

Formic acid, introduced to replace earlier Scandinavian additives that were based on hydrochloric acid and sulphuric acid, is well established as an effective chemical for improving the quality of silage (see Ch. 6). It is easy to apply and the quantity required per tonne of fresh crop is not large. But formic acid is a volatile chemical and, in an attempt to reduce handling problems and increase user comfort, the acid salt ammonium tetraformate has been introduced. Laboratory tests suggest that this salt is very nearly as effective as the acid in preventing secondary fermentations.

Formalin, used alone as a silage additive, can induce secondary fermentations if added at levels of less than 5 kg/t fresh crop, because it only partially restricts the fermentation and allows the development of clostridial bacteria rather than the lactobacilli. Thus its use is restricted to mixtures with acids. Often, however, recommended rates of use of commercial mixtures are too low; there is insufficient formalin to protect protein from degradation and there is insufficient acid to prevent secondary fermentation. The recommended rates of addition shown in Table 3.4 indicate that the proportion of acid to formalin should be about 2 to 1. At 1 litre of formalin per tonne, the level of addition of formaldehyde is about 1 per cent of the crude protein in the DM.

Urea is commonly used to increase the nitrogen content of grasses that have low contents of nitrogen, such as maize. At the recommended rate, the level of crude protein equivalent (nitrogen $\times$ 6.25) should be increased by about 8 percentage units (from 8 to 16% in a crop of forage maize harvested at 35 per cent DM). Often urea is added either with molasses (to improve the fermentation of tropical grass crops) or minerals (which tend to be lower in grasses such as maize and other sub-tropical species than in the younger temperate species and the legumes). Solutions may be added, but powders can also be used, provided they are carefully metered on to the crop to ensure even distribution. Mixtures of ammonia, molasses and minerals have also been successfully used with forage maize in north America and eastern Europe, as has anhydrous ammonia alone. In the latter case the chemical may be introduced to the crop at the point of entry to the silo, via an expansion chamber that allows the gas to vaporize (the 'cold flow' method). In this way the loss of ammonia to the atmosphere is minimized and, at the same time, a good mixing with the crop can be achieved. The recommended rate of addition of ammonia (as anhydrous) to maize forage is 5.5 kg/t fresh crop weight.

Losses during ensilage

Typical losses of DM associated with the ensiling of grass in bunker silos under good management conditions are shown in Table 3.5.

Losses of DM in the field are difficult to estimate but, clearly, they are likely to be greater with wilted crops than with direct-cut material. It has been estimated that the loss of DM during wilting averages 1.5 per cent per 24 hours in the field under European conditions.

Table 3.5   Losses likely to occur from direct-cut or wilted grass ensiled in bunker silos under conditions of good management

DM loss (%)	Direct-cut*	Wilted†
In field		
Respiration	—	2
Mechanical losses	1	4
During storage		
Respiration	—	1
Fermentation	5	5
Effluent	6	—
Surface waste	4	6
During removal from store	3	3
Total (%)	19	21

* Formic acid added at 2.5 litres/t
† 36-hour wilt in the field

The loss of DM due to fermentation depends very largely on the pattern that occurs. Fermentation of glucose and fructose to lactic acid by the streptococci and lactobacilli is an extremely efficient process, particularly in terms of energy conservation. By contrast, the secondary fermentation of lactic acid to butyric acid is associated with substantial losses of both DM and energy.

Since more DM is usually lost during fermentation than energy, there is often an increase in the concentration of gross energy in silage compared with that in the ingoing crop. This partially alleviates the loss of DM, which probably amounts to about 5 per cent of the DM for crops that do not undergo a secondary fermentation (Table 3.5.).

Despite low losses in the field with direct-cut crops, this is offset by the loss that commonly occurs as a result of effluent production during storage. Studies of the relationship between DM content of the ensiled crop and effluent production show that, with bunker silos, effluent is unlikely to occur if the crop is wilted to a DM content greater than 25 per cent (35 per cent for crops ensiled in towers). Below this level, large volumes of effluent can be produced soon after ensiling and, in the case of tower silos, the hydraulic pressures built up towards the base of the structure can constitute a serious risk to stability.

Silage effluent is some 20 times more pollutant to rivers and streams than is cattle manure. Thus its production should be contained and disposal should occur after dilution with water (1:1) on land well away from rivers.

Losses of DM due to surface waste can be very high if the silo has been incompletely sealed. As much as 30 to 40 per cent of the material may be rejected as unfit for consumption by animals and, clearly, this represents an intolerably serious source of wastage.

Complete sealing of the silo is essential in order to reduce loss due to surface waste; with adequate protection of the seal from damage by wind, animals and birds, losses from this source can be contained at a low level. Under conditions of good management surface waste should not exceed 6 per cent of the DM ensiled.

Aerobic deterioration at the silo face during the feed-out period can pose a problem, particularly in hot weather and when the density of the material in the silo is relatively low. Penetration of air into the mass of the silage should be discouraged, as far as is possible, by uniformly removing the crop and not leaving large amounts of disturbed material lying around at the silo face. This source of loss can be minimized in the case of trough-feeding by the use of specialized unloading equipment and by designing bunker silos with a narrow face, which is removed frequently.

With total losses of DM from cutting to feeding of less than 25 per cent, there should be only a small decrease in the digestibility of the conserved product compared with that of the ingoing crop. Well preserved wilted silages are commonly about 2 percentage units lower in digestibility of organic matter than the comparable ingoing crop and, under experimental conditions, well preserved direct-cut crops have shown similar values for digestibility compared with the crop at the time of ensiling.

Crops that undergo severe secondary fermentation may, however, lose up to 8 percentage units of digestibility, associated with the relatively high loss of components that are completely digestible.

Other crops
Legumes

Reference has already been made to the conservation of legumes, in the context of ensilage. Hay can be made from legumes; lucerne (alfalfa) is commonly conserved in this way in central Europe, north American and Africa. A major problem with making hay from legume crops is the loss of leaf due to the action of the haymaking machinery, during turning in the later stages of field drying, and at baling. Thus ensilage is preferred, although the lower content of WSC in legumes means that an additive is essential.

Forage maize

Forage maize is a particularly attractive crop for conservation, and for giving to beef cattle. Harvested in late summer as a mature crop, yield of DM should average 10 t/ha (Fig. 3.3). At harvest, the crop has a relatively high concentration of metabolizable energy (10.7 MJ/kg DM), largely reflecting the significant contribution of the ear to total yield (50%). The target DM content at harvest is 25 per cent (see Table 4.1), although the crop is often harvested at higher levels of DM, particularly in north America where it is commonly ensiled in towers. Although relatively high in the content of starch (in the ear), the content of WSC is relatively high and the buffering capacity of the crop relatively low, so that the material is almost invariably well preserved when ensiled, with a predominantly lactic acid fermentation. The content of nitrogen in forage maize is relatively low compared with that in other grasses (see Table 4.3) and supplementary nitrogen is normally required in diets for beef cattle. The use of forage maize for beef production is discussed in Chapter 6.

Whole-crop cereals

Whole-crop cereals (rye, wheat, barley, oats and sorghum) are often ensiled as a tactical measure when yields of grass or of the cereal-grain are inadequate. When harvested at the milk-ripe stage of grain maturity there is risk of poor preservation and the optimum time for harvest of these crops appears to be at the dough-ripe stage when the whole crop DM has reached at least 40 per cent. Fortunately, this stage usually coincides with maximum yield of DM per hectare.

Other annual forages

Forage crops such as kale and rape are rarely conserved; they suffer from a low content of DM, and precautions should therefore be taken if they are ensiled to avoid poor preservation and excessive effluent production.

By-products

By-products are often conserved for use as feed for beef cattle (see Ch. 7). Materials such as sugar beet tops tend to be relatively low in DM content and should therefore be ensiled with an additive, preferably on top of a drier crop or straw, to absorb effluent.

Straws and maize stover (stems, leaves, cob and husks) are usually stored as bales, but they can also be ensiled. In the latter case they are often treated with an additive such as sodium hydroxide to improve digestibility. The DM content of the treated materials should not be less than 60 per cent or the risk of inducing a clostridial fermentation will be greatly increased.

Potatoes may be conserved for use as a feed by piling them up, to make an unwalled clamp, and covering the clamp with straw. It is advisable to put soil or manure on top of the straw to keep the straw in close contact with the potatoes. This double covering protects the tubers from frost damage. If possible, the clamp should be sited on concrete to avoid contaminating the potatoes with soil when they are removed from the clamp.

Chapter 4 Evaluation of conserved forages

The production of beef from conserved forages involves the formulation of diets to give certain levels of intake of metabolizable energy, to enable targets for live-weight gain to be reached. In order to make a realistic assessment of the potential of the forage to supply the required nutrients to the animal, it is essential to know two major attributes: the nutrient composition of the material; and its likely level of consumption by the animal when given alone or in a mixed diet.

Composition   Historically, the composition of feeds was assessed in the laboratory in terms of dry matter (DM), nitrogen (crude protein), ash, fibre, fat (ether extract) and carbohydrates other than fibre (by difference). Estimates were then made of energy value, as starch equivalent, and diets were formulated accordingly.

We are now in the position of knowing much more about the interaction between laboratory measures of composition and the response of the animal in terms of intake or digestibility. Thus the evaluation procedures described here concentrate on those determinations that can assist the beef producer in formulating diets based on conserved forages for specific animal responses – intake and weight gain.

Dry matter   DM *per se* appears to have relatively little effect on the value to the animal of conserved forages, but there are other important factors closely linked to DM, such as stability during storage. Thus it is essential to be able to measure DM rapidly and accurately, particularly when crops are being harvested.

Unfortunately, the measurement of DM on the farm is not

easy. If an oven is not available for drying samples, oil may be used to remove water from the plant material. A known weight of crop is chopped and placed in a pan containing oil (also of known weight) and heated until 'boiling' occurs, indicating that water is being driven off. The temperature should not exceed 115 °C (it is useful to have a thermometer in the pan) to avoid removing components other than water. When 'boiling' has ceased and no more bubbles are produced, the pan and its remaining contents are re-weighed. The difference in weight is assumed to be the water content of the crop.

Recommended DM contents for harvesting crops for conservation are given in Table 4.1. For the production of hay with barn drying or addition of chemical preservative, it is necessary to harvest the crop at between 65 and 75 per cent DM. With field-dried crops, 80 per cent DM is necessary for safe storage, although care must be taken with legumes to avoid excessive loss of leaf due to the action of machinery on the crop during the later stages of drying.

Higher DM contents are recommended for tower silos than for other types of silo because it is necessary to avoid effluent production from these structures. Very high hydraulic pressures can be created towards the bottom of tower silos filled with crops of less than 30 per cent DM content and these pressures can result in the collapse of the tower.

Table 4.1   Recommended contents of crop dry matter at harvest for conservation (%)

Crop	Hay		Silage	
	For barn drying or with preservatives	Field dried	Bunker or clamp silos	Tower silos
Perennial grasses Italian ryegrass	65–75	80	20–30	35–55
Whole-crop cereals	–	–	40–50	
Legumes	65–75	80	25–35	35–55
Maize, sorghum	–	–	25–35	30–35

The recommended levels of DM for legumes and for whole-crop cereals are higher than for other crops, to reduce the risk of secondary fermentation in the silo. In the case of legumes such as lucerne (alfalfa), an effective addictive should also be used.

Ash

The content of ash indicates whether or not soil has contaminated the conserved forage; if values for total ash exceed 12 per cent of the DM, then it is most likely that the material is contaminated. Intake may be reduced in consequence and, occasionally, a secondary fermentation can be induced by the presence of soil in ensiled crops.

Crops with relatively high levels of ash, such as the legumes, tend as a result to be relatively rich in the major mineral elements required by ruminant animals. It is safer, however, to assume that beef cattle will always require supplementary minerals in the diet and the manufacturer's recommendations should be followed when adding a proprietary mineral mixture to the ration.

Nitrogen

The N content of conserved forages is a most important constituent, and can provide a useful indication of the quality of the product and the conditions under which it has been stored.

Total N, multiplied by 6.25 gives the content of *crude protein* (CP) in the material. In the case of silages, this determination should be carried out on samples of the fresh silage, since there are nitrogenous components, such as amines, amides and ammonia, that can be lost from the material during drying. Indeed, an approximate indication of the proportion of volatile (non-protein) N can be obtained by determining N on samples of both fresh and dried material. The difference between the two values is the N lost during drying (i.e. the volatile fraction). Silages with a high proportion of *volatile nitrogen* are likely to have undergone secondary fermentation in the silo and are considered to be poorly preserved (Fig. 4.1).

A more common measure of secondary fermentation is the proportion of total N present in silages as *ammonia nitrogen*. This is measured directly on extracts of fresh material. Well preserved silages usually contain less than 10 per cent of total N as ammonia-N (see Table 4.2). Ammonia-N is the most useful single measure of the quality of preservation but, because secondary fermentations develop only slowly during the storage

Figure 4.1   The nitrogenous components of fresh grass, well preserved silage and poorly preserved silage

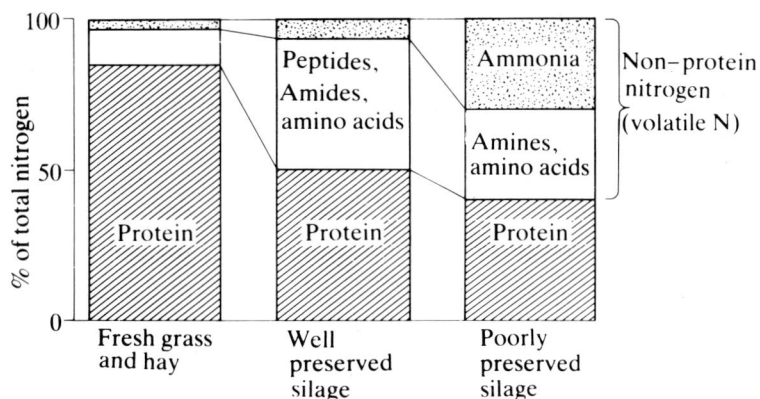

Table 4.2   *Nitrogenous components in conserved forages: typical values for well preserved and poorly preserved hay and silage (% of total nitrogen)*

	Well preserved	Poorly preserved
Hay		
Hot water-insoluble N	85	65–75
Acid detergent-insoluble N	<10	>15
Silage		
Hot water-insoluble N	50	35–45
Volatile N	<15	>20
Ammonia-N	<10	>15
Acid detergent-insoluble N	<5	>10

period, it should be determined after the crop has been ensiled for at least 120 days.

The *degradability* of the total CP fraction represents the proportion of feed protein likely to be degraded in the rumen. Silages contain a very high proportion of degradable protein and degradability values in excess of 0.9 are often found. The protein in hays may also be substantially degradable (0.8–0.9), whilst artificially dried forages contain much less degradable protein as a result of the heat-treatment of the protein fraction during drying. The degradability of the protein in artificially dried grasses and legumes is likely to be between 0.4 and 0.6. Formaldehyde-treated silages may also have a relatively low content of degradable protein. In the laboratory, the proportion of degradable protein is determined by incubation of samples of feed in a buffer solution, which simulates conditions in the rumen. The soluble fraction is considered to be degradable in rumen fluid.

An approximate measure of the proportion of 'true' protein in conserved forages in obtained by determining the *nitrogen insoluble in hot water*; this fraction may be degraded in the rumen by microbial enzymes to non-protein nitrogenous (NPN) compounds, but a proportion will escape degradation. It is a measure of the N that is likely to be of most value to the animal, since the remainder (the soluble fraction) is likely to be rapidly absorbed from the rumen without being captured by the rumen microorganisms. When this happens, it is of no use to the animal and is excreted as urea via the urine.

By contrast, some protein can become complexed with sugars in Maillard-type reactions during the storage period. These reactions occur at temperatures in excess of 35 °C and the products are almost completely indigestible by the animal. Hays and high DM silages that have heated excessively during storage can contain quite high proportions of their total N in this heat-damaged form. It can be measured by determining the proportion of total N that is *insoluble in acid detergent* (see p. 00). This fraction should not exceed 10 per cent of the total N in well preserved conserved forages (Table 4.2).

Since heat-damaged protein is unavailable to the animal, there is a relationship between acid detergent-insoluble N and digestibility of N (see p. 00).

The fermentation process in the silo is reflected in the degradation of a proportion of the plant protein to (water soluble) volatile nitrogenous compounds. Commonly, silages contain about half their total N as hot water-insoluble N; the remainder comprises amino acids, amides, amines and ammonia in different proportions, depending on the pattern of fermentation (Fig. 4.1). Lactic acid fermentations are reflected in low amounts of amine-N and ammonia-N in the product, whilst clostridial (secondary) fermentations result in deamination of amino acids, with a corresponding increase in the amine and ammonia-N. The carbon skeletons of the amino acids remain as acetic acid in the silage or are lost as carbon dioxide.

Table 4.3 Typical values (and range) for the nitrogen content of conserved forages

	Total N (% of dry matter)*	
	Temperate crops	Tropical crops
Hay		
Grass	1.6 (1.3–2.1)	1.2 (0.8–1.8)
Legume	2.8 (2.2–3.6)	2.4 (2.0–2.8)
Silage		
Grass	2.4 (1.6–2.7)	1.8 (1.2–2.0)
Legume	3.2 (2.6–3.6)	2.6 (2.2–2.9)
Whole-crop cereal	1.6 (1.2–1.8)	–
Maize, sorghum	1.4 (1.2–1.6)	1.1 (0.8–1.3)

* For crude protein content, multiply by 6.25

Legumes are typically much higher in N than grasses; annual grasses (maize, sorghum, wheat, barley, oats and rye), which tend to be harvested for conservation after flowering when the crop is physiologically mature, contain lower contents of N than do perennial grasses. Tropical grasses contain lower contents of N than do temperate grasses. The range in contents of N within crops, given in Table 4.3, reflects both stage of growth at harvest and the level of fertilizer-N applied to the crop. With high levels of fertilizer there can be luxury uptake by the plant, and this can result in nitrate and nitrite-N being present in the crop at harvest. There is no evidence, however of nitrates or nitrites in conserved forages creating serious problems when the material is given to cattle. Hays, normally conserved at a more advanced stage of growth and grown on land that has received less N, usually contain lower contents of N than silages (Table 4.3). Exceptions to this generalization are silages made from whole-crop cereals and maize, which contain relatively high contents of starch in the grain fraction, and which are normally harvested at an advanced stage of crop maturity. As a result, diets containing these crops require supplementation with additional N (see Ch. 6).

Fibre   The fibrous constituents of plant material are the cell walls, which give the plant its structure.

Crude fibre is a useful measure of the stage of growth at which the grass was harvested and is commonly used in the prediction of digestibility or energy value.

A more useful measure, related not only to digestibility but also to the voluntary intake of forages, is the content of *neutral-detergent fibre*. This is determined by boiling a sample of plant material in detergent solution at neutral pH. The residue is the cell wall fraction, which is only partially digested by the animal. It comprises principally hemicellulose, cellulose and lignin, and the extent to which these components are digested depends largely on the amount of lignin, which accumulates in the cell walls as the plant matures and gives them their structural rigidity. Lignin itself is extremely stable and almost completely indigestible. It encrusts the other components of the cell wall and forms chemical bonds with cellulose, which prevent the cell wall matrix from swelling – a prerequisite for the penetration of microbial enzymes. Thus the result of lignification is that parts of the cell wall fraction are rendered unavailable to microbial digestion in the rumen.

The procedure for determining the content of cellulose and lignin in forages is to reflux the material in acid detergent. The residue, the *acid-detergent fibre*, is a useful laboratory measure, since it is related to the digestibility of the crop (see below).

The difference between the content of neutral-detergent fibre and acid-detergent fibre is the hemicellulose in the material.

The residue of acid-detergent fibre can be used to determine the content of *acid-detergent lignin* by treating it with concentrated sulphuric acid. The total N content may also be determined, to provide an indication of the extent to which the protein in the crop has been damaged by heat (see p. 62).

The difference between acid-detergent fibre and acid-detergent lignin is the content of cellulose in the plant material, but this complete cell wall analysis is not normally carried out on samples of conserved forages from farms. Generally, the analysis is restricted to acid-detergent fibre, from which digestibility or energy value of the crop is predicted.

pH value   This relatively simple determination is of use in indicating the pattern of fermentation in silages. Silages normally undergo fermentation to reach a stable pH, depending on their content of DM (Fig. 3.16). Thus a wilted crop of 30 per cent DM content would be expected to reach a stable pH of about 4.5, whilst well preserved direct-cut grasses, including maize and sorghum, would be expected to give lactic acid-dominant fermentations, to produce silages with pH values between 3.8 and 4.2.

By contrast, direct-cut crops that have undergone secondary fermentations generally have pH values above 4.5 since, below this pH, lactic acid bacteria tend to dominate to the exclusion of clostridia. Thus a simple test of fermentation pattern is to determine pH. If the crop is wet and has a pH above 4.5, then in all probability some secondary fermentation has occurred. If the pH is relatively low, all is well.

Litmus paper is available to test the pH value of silage juice on farms within quite narrow ranges; this is particularly useful if laboratory analysis is not available.

Fermentation acids

Some laboratories determine, in addition to pH, the content of fermentation acids in silages (principally lactic, acetic, propionic and butyric acids). This analysis gives a complete picture of fermentation pattern and indicates whether or not a secondary fermentation has occured.

As a guide to silage quality, targets have been set for the composition of well preserved silage (which is likely to be eaten in quantities similar to the fresh or artificially dried forage). The targets are given in Table 4.4.

These targets can best be met by wilting the crop prior to harvest and by the use of an effective additive to reduce the extent of fermentation.

Table 4.4 Targets for the composition of well preserved silage

Acetic acid (% of DM)	≤2.5
Other volatile fatty acids	None
Hot water-insoluble N (% of total N)	>50
Ammonia-N (% of total N)	≤5

Voluntary intake

Since beef cattle are usually given conserved forages *ad libitum*, it is important to know how much of a particular feed the animal is likely to eat in order to predict its rate of growth or determine the level of supplementation.

The major factors affecting the amount of DM eaten are the live weight of the animal, the digestibility of the conserved forage and, in the case of silages, the fermentation in the silo and the particle length of the material. A further important effect on forage intake is the level of supplementation, since with most forages there is a decrease in intake with increasing amount of supplementary feed in the diet.

Although heavier animals eat greater quantities of conserved forage DM than lighter animals, intake relative to live weight tends to decrease with increasing weight. Intake is therefore often expressed per unit of live weight, particularly when comparisons between different forages are made.

The intake of conserved forages generally increases with increased digestibility, as shown in Fig. 4.2 for hay. However, the relationship for silage is less clear than that for hay. But, even with hays, other factors (preservation quality and species of crop) can influence the relationship; thus there are differences in intake between species of similar digestibility. The most striking difference is that between legumes and grasses, with the intake of the legume usually being markedly higher than that of grass hay (see Ch. 5).

Figure 4.2   Effect of digestibility of ryegrass hay on voluntary intake by young cattle. From Tayler, J. C. and Lonsdale, C. R. (1970) EAAP, Godollo

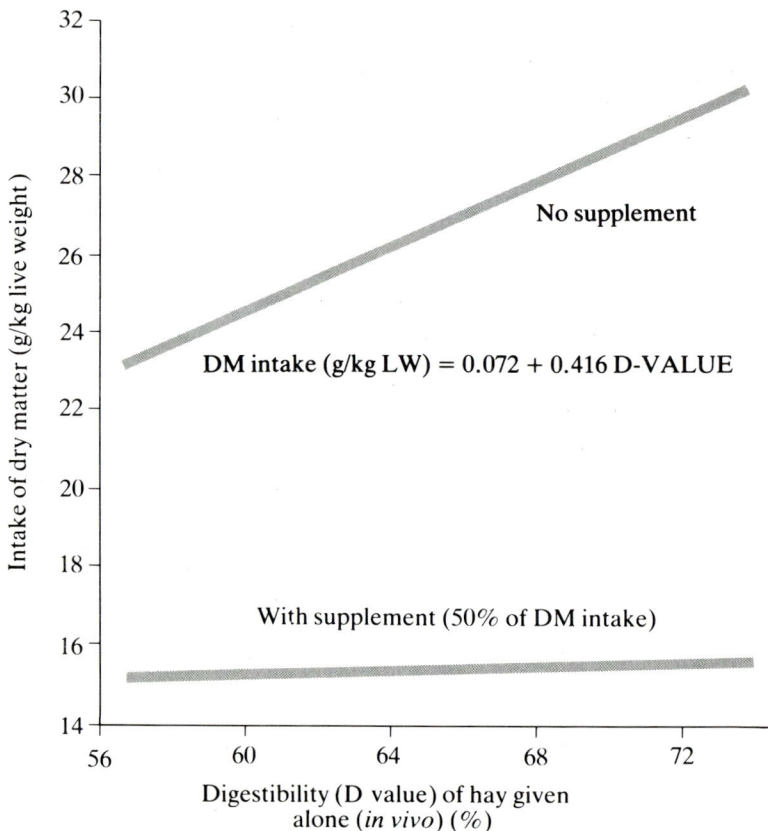

No supplement

DM intake (g/kg LW) = 0.072 + 0.416 D-VALUE

With supplement (50% of DM intake)

Intake of dry matter (g/kg live weight)

Digestibility (D value) of hay given alone (*in vivo*) (%)

With well preserved silages, the following relationship between intake of DM by beef cattle given silage as the sole feed, and digestibility and silage composition, was derived from 37 experiments (from Flynn, A. Y. 1981. In *Recent Advances in Animal Nutrition–1981*. Butterworth, London):

$$DMI = 0.953 + 0.019DMD - 0.187pH + 0.015DM$$
$$r^2 = 0.50; \text{ s.e.} = 0.13,$$

where DMI = dry matter intake as a percentage of live weight;
  DMD = dry matter digestibility; and
  DM = dry matter content of silage (%).

It has long been considered correctly, that the pattern of fermentation overrides any relationship that may exist between intake and digestibility in silages. Even with well preserved silages, intake is positively related to the content of DM and negatively related to the pH value of the silage, in addition to being positively related to digestibility. A silage of 76 per cent DMD, 20 DM and with a pH of 3.9 would be expected from the above equation to result in a DM intake of 2.0 per cent of live weight.

The effect of fermentation on voluntary intake is shown in Tables 4.5 and 4.6. In the first example (Table 4.5), intake of the silage that fermented to a lesser extent was greater than that that was more extensively fermented, despite the fact that the latter silage was of higher digestibility. The net effect was that intake of digestible organic matter was very similar between the two silages. In the second example (Table 4.6), the same crop gave a poorly preserved product when ensiled without additive, whilst that made with addition of formic acid (2.3 litres/t) was well

Table 4.5 Effect of extent of fermentation on intake of silage

	Extent of fermentation	
	More	Less
DM (%)	20	26
pH	4.0	4.3
Total fermentation acids (% of DM)	10.0	6.5
Digestibility-value *in vitro*	64	58
Intake of silage (kg DM per day)	7.3	8.0

Source: from Tayler, J. C. and Aston, K. (1976) *Animal Production* **23**:211

Table 4.6   Effect of pattern of fermentation on intake of silage by beef cattle

	Preservation	
	Good	Bad
pH of silage	4.2	4.8
Ammonia-N (% of total N)	7	18
Digestibility of DM (% of live weight)	74	71
Voluntary intake of DM (% of live weight)	1.9	1.4
Carcass weight gain (kg carcass per day)	0.50	0.34

Source: from Flynn, A. V. (1981) In *Recent Advances in Animal Nutrition – 1981*. Butterworth, London

preserved. Intake of DM of the well preserved silage was some 36 per cent greater, and intake of digestible DM some 41 greater than that of the poorly preserved product.

Short-chopping of the crop can increase the voluntary intake of silage DM (by up to 20% in cattle),partly by improving the quality of preservation, but also through an increase in the speed of eating, and a reduction in the period of time between the end of a meal and the start of rumination. Crops should therefore be chopped quite short prior to ensilage and a target average particle length of 25 mm should be sufficiently short to overcome the problem of the matting of particles in the rumen, which can prevent regurgitation of feed boli during rumination. Very short chopping, although beneficial to the fermentation by facilitating rapid release of water-soluble carbohydrates in the silo and to intake, is associated with a much greater consumption of fuel during harvesting. Thus the recommended average particle length of 25 mm represents a compromise between high fuel consumption and reduced nutritive value of the product.

Prediction of intake in mixed diets

Thus far we have concentrated on intake when conserved forages are given to cattle as the sole feed. It is essential in formulating diets to meet performance targets to know: first, the likely level of intake of the conserved forage when given alone; and secondly, the extent to which that intake is likely to be reduced when concentrates are included in the diet. The values in Table 4.7 represent estimates of the likely levels of consumption of silages

Table 4.7   Probable values for the intake of dry matter of hays and silages when given as the sole feed to growing beef cattle (kg/day)

Live weight (kg)	Early cut (10 MJ metabolizable energy (ME) per kg DM)					Medium (9 MJ ME per kg DM)					Late (8 MJ ME per kg DM)					Maize silage[†]
			Silage*					Silage*					Silage*			
	Hay	1	2	3	4	Hay	1	2	3	4	Hay	1	2	3	4	
100	2.7	2.4	2.0	1.8	1.4	2.4	2.2	1.9	1.7	1.3	2.2	1.8	1.6	1.4	1.2	2.3
200	4.6	4.4	3.8	3.4	2.6	4.0	3.6	3.4	3.0	2.4	3.6	3.2	3.0	2.8	2.2	4.6
300	6.0	5.6	5.2	4.6	3.6	5.4	4.9	4.6	4.1	3.2	4.8	4.4	4.1	3.6	2.8	6.4
400	7.2	6.8	6.2	5.4	4.4	6.4	5.8	5.4	4.8	3.8	5.6	5.0	4.7	4.2	3.3	8.0
500	8.0	7.4	6.8	6.0	4.8	7.0	6.3	6.0	5.3	4.2	6.0	5.4	5.1	4.5	3.5	8.9
600	8.4	7.8	7.4	6.4	5.2	7.2	6.5	6.1	5.4	4.3	6.1	5.5	5.2	4.6	3.6	9.3

* Grass silages:
  1. wilted, more than 30% DM;
  2. additive-treated (formic acid or formalin + acid);
  3. Well preserved direct-cut pH 4.0;
  4. Poorly preserved

[†] Maize silage + 1.5 kg per day protein supplement

and hays of differing quality when given as the sole feed to beef cattle of different live weights. Those quoted for maize silage presume that a supplement is given to rectify the low content of N in the material.

The probable substitution values for hays and for silages of differing quality are in Fig. 4.3. and 4.4. respectively. The extent to which intake is reduced by concentrates depends to a large extent on the intake of the forage when given as the sole feed. This is a reflexion of digestibility in the case of hay, whilst with silage the fermentation is also important, as has already been seen (Tables 4.5 and 4.6). In the case of maize silage, the appropriate substitution value is 0.8.

An alternative approach is to use equations to predict intake. In the case of hays, the equation below (from Marsden's (1983) Ph. D thesis) was derived from 102 sets of data in which hay was given *ad libitum* to beef cattle either alone or with concentrates:

$$I = 0.0421DOMD + 2.174C - 0.00171LW - (0.0299DOMD)C,$$

Figure 4.3   Substitution values for hay when given to beef cattle with concentrates.
From Marsden, (1984) *Ph. D. Thesis, University of Reading*

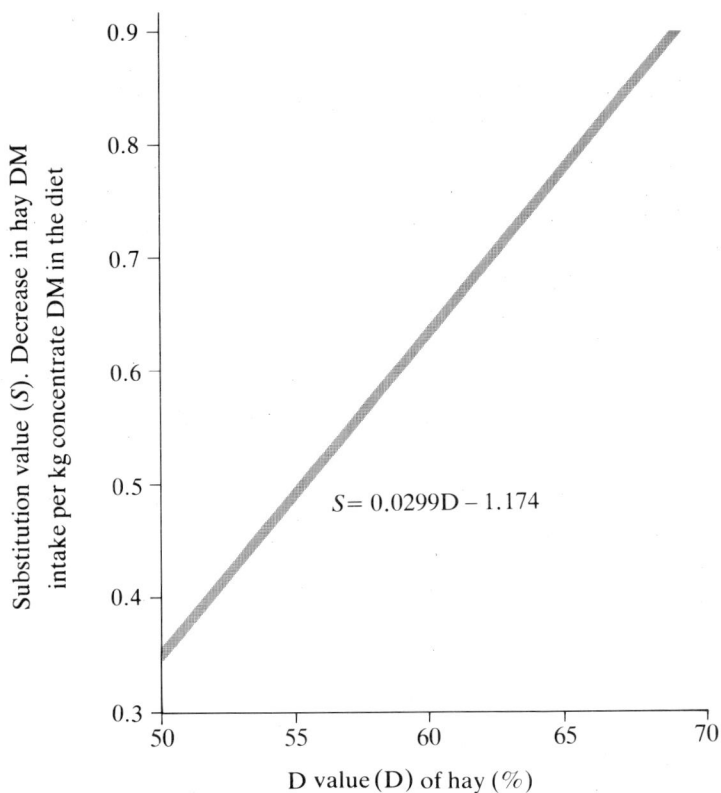

where I = intake of DM (kg/100 kg LW);

DOMD = D-value of hay (%);

C = intake of concentrates (kg per 100 kg LW); and

LW = live weight (kg)

For silage, a two-stage approach has been developed at the Edinburgh School of Agriculture by Dr M. Lewis and reported in the proceedings of the 1981 6th Silage Conference held there. The first stage comprises a prediction of the intake of the silage when given alone, from a knowledge of its composition and the live weight of the animal (A). This intake is then corrected for the substitution effect of concentrates (B). The equations are as follows.

Figure 4.4 Probable substitution values for silage when given to beef cattle with concentrates. When silage is of low digestibility, substitution values are also low, especially with poorly preserved crops. The values for substitution are higher for well preserved than for poorly preserved silages at all levels of digestibility

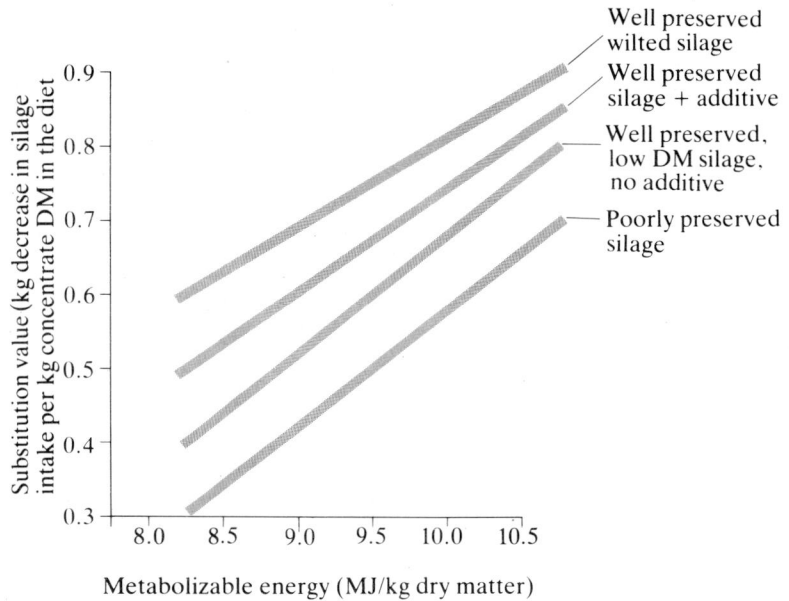

Metabolizable energy (MJ/kg dry matter)

A   Intake of silage when given alone (I) to:
   (i) weaned suckled calves
      $I = 0.0105DM + 0.0156D + 0.0075LW - 0.02NH_3 + 3.5$;
   (ii) artificially-reared calves
      $I = 0.010DM + 0.0161D - 0.0154LW - 0.02NH_3 + 13.6$.

B   Intake of silage in mixed diets (IMD):
      $IMD = 0.92I - 0.027I \times C - 0.0247C^2 + 1.0$,
   where   DM = Dry-matter content of silage (g/kg, maximum 350);
      D = digestible organic matter in the dry matter (DOMD) of silage (g/kg);
      NH_3 = ammonia-N in total N (g/kg, maximum 250);

C = intake of concentrate DM (g/kg LW);
LW = live weight of animal (kg); and
I = intake of DM (g/kg LW).

Digestibility The term digestibility describes that proportion of the feed which provides useful energy or protein to the animal. In Europe it is normally expressed on the basis of either DM or organic matter (OM), and is commonly expressed as the *content of digestible organic matter in the dry matter* (*DOMD or D-value*). In the USA and in some other countries, digestibility is often expressed as the percentage of *total digestible nutrients* (TDN) in the DM. With conserved forages, digestibility describes the stage of growth of the crop at harvest (see Ch. 3) and, as we have already seen, it interacts with the level of concentrate supplement to determine the total feed intake of the diet.

The proportion of the feed that is apparently digested by the animal is measured as the difference between the amount consumed and the amount voided in faeces, expressed as a proportion of the amount consumed. The term 'apparent' digestibility is used because the calculation does not take into account the excretion of endogenous material in faeces (such as metabolic secretions into the alimentary tract and epithelial cells from the intestines sloughed off by the action of digesta).

Thus dry matter digestibility (DMD) % =
$$\frac{\text{DM consumed} - \text{DM in faeces}}{\text{DM consumed}} \times 100;$$

organic matter digestibility (OMD) % =
$$\frac{\text{OM consumed} - \text{OM in faeces}}{\text{OM consumed}} \times 100;$$

and digestible organic matter in the dry matter (DOMD or D-value) % =
$$\frac{\text{OM consumed} - \text{OM in faeces}}{\text{DM consumed}} \times 100$$

or $\dfrac{\text{OMD\% } (100 - \text{ash \%})}{100}$

Determination *in vitro* A useful laboratory procedure is to simulate the digestive process by incubating dried and ground samples of feed, first with rumen liquor for 48 hours from an animal given a similar diet, then in acid-pepsin for 48 hours to simulate the action of the abomasum. The insoluble residue is the indigestible DM, which can be ashed to determine the indigestible OM and thus D-value *in vitro*. The technique has been widely adopted and is particularly useful for fresh forages and hays. With silages, however, the problem of loss of volatile compounds during sample preparation can lead to underestimation of digestibility. Freeze-drying of the material prior to incubation can help to alleviate the problem.

With by-products and crops of very low N content it is necessary to add a small quantity of supplementary N (as urea or ammonium sulphate) to the first stage of incubation.

Prediction of digestibility and energy value

Most laboratories are equipped for the determination of digestibility *in vitro* but the technique is rather time-consuming, and efforts have been made to simplify the procedure by using more rapid and simple determinations of such constituents (for instance fibre), to predict digestibility and the content of metabolizable energy in the material. The following equations devised by Mr P. Barber and his co-workers in MAFF/ADAS were in use in the UK in 1982.

For hays:
$$DOMD = 102.3 - 0.120 \, MADF;$$
$$ME = 0.017 \, DOMD \, (\textit{in vitro}) - 1.1;$$
$$ME = 16.5 - 0.021 \, MADF.$$

For silages:
$$ME = 2.16 + 0.0186 \, DOMD \, (\textit{in vitro}) + 0.0128 \, CP - 0.027 \, DM;$$
$$ME = 14.6 + 0.0075 \, CP - 0.0123 \, MADF - 0.0030 \, DM,$$

where DOMD = content of digestible organic matter in the dry matter (g/kg);
ME = content of metabolizable energy in the DM (MJ/kg DM);
MADF = modified acid-detergent fibre (g/kg DM);
CP = crude protein (g/kg DM); and
DM = dry matter (g/kg fresh weight).

Chapter 5 Beef from dried forages

In this chapter the performance of beef cattle given hays and artificially dried forages is considered, and some systems of production based on these feeds are described. The levels of intake and rates of live-weight gain to be expected from beef cattle given dried forages are compared with those from silage made from the same initial crop; effects of the quality of the dried forage and level of concentrate supplementation on performance are also considered. The growth of beef cattle given hay treated with preservative is compared with that from barn-dried hay. Some alternative crops to grass are discussed, with particular reference to dried legumes and dried forage maize.

Dried forage compared with silage

Many farmers make both hay and silage, often with the intention of giving hay to the younger cattle and silage to the older animals. This practice would appear to be nutritionally sound, since there is evidence, at least with artificially dried grass, that calves given the dried product can achieve substantially higher rates of live-weight gain than those given silage made from the same initial crop (Table 5.1). The most probable explanation of this difference in rate of growth is that the calves (100 kg live weight) given dried grass had a higher efficiency of utilization of nitrogen than those given silage, so that lean tissue gain was increased. It is notable that the levels of intake and digestibility were similar, but that the proportion of insoluble (protein) N was markedly higher in the dried grass than in the silage. This may well have led to an enhancement in the supply of undegraded dietary protein to the body tissues (see Ch. 2).

Thus dried forages have a particularly useful role in the feeding of the young beef animal, whilst silages appear better suited to the finishing period, when requirements for protein are lower.

Table 5.1   Performance of young beef cattle given dried grass or silage of high quality, made from the same initial crop

	Silage*	Dried grass
Composition		
Dry Matter (DM) (%)	29	86
Acid-detergent fibre (% of DM)	22	24
Ash (% of DM)	7.6	6.3
Nitrogen (% of DM)	3.5	3.4
Hot water insoluble[†] N (% total N)	28	79
pH	4.0	–
Digestibility (D)-value *in vivo* (%)	70	69
Voluntary intake		
DM (g/kg live weight (LW))	24	26
ME (MJ/100 kg LW)	28	27
LW gain (g/day)	370	760

* Made with the addition of formic acid at 2.2 l/t fresh crop
[†] i.e. protein N rather than non-protein N

But young cattle also need energy in order to gain in weight and, since the intake of dried forages increases with increasing digestibility, there is a marked effect of the quality of dried forages on liveweight gain.

Quality of dried forages and beef cattle growth

With the increase in intake of dried forage as digestibility increases (Fig. 4.2), there is therefore a substantial increase in the supply of digestible nutrients to the animal. The effect of this on rate of live-weight gain is shown in Fig. 5.1.

In general, live-weight gains increased by about 25 g/day for each percentage unit increase in digestibility (D)-value. Thus, with early-out material of high quality (D-value 69%, metabolizable energy (ME) 10.3 MJ/kg dry matter (DM)), gains close to 1.0 kg/day should be achieved from dried forage given *ad libitum* as the sole feed (Fig. 5.1).

There appears to be a similar response in weight gain to the feeding of concentrates with dried grass, at least with young beef cattle. When concentrates comprise about half of the diet – 2 kg DM per head per day or so – weight gains can be improved by as much as 400 g/day. Calves gaining 600 g/day on grass alone can

Figure 5.1   Performance of beef cattle given dried forage of different qualities. As forage quality increases then the dual effects of an increase in nutrient concentration and an increase in feed intake work together to produce substantial increases in the rate of live-weight gain

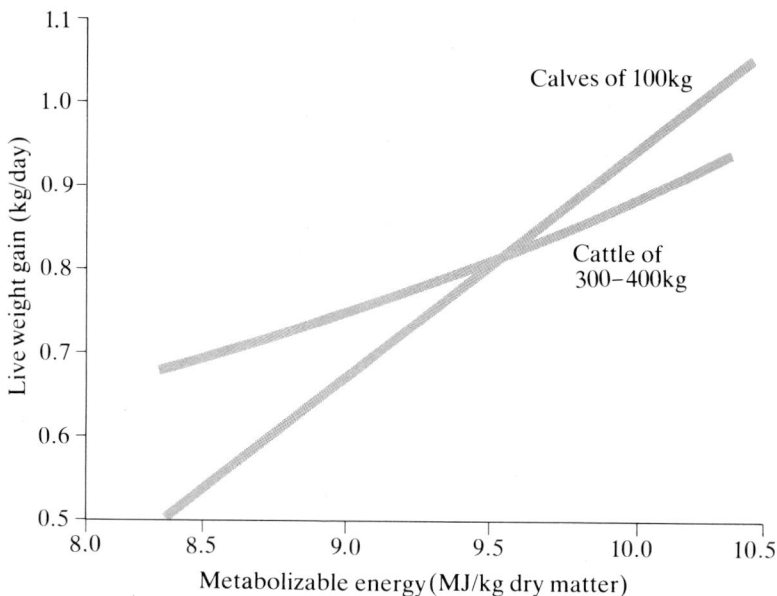

therefore gain 1.0 kg/day on grass plus concentrates. Calves gaining 800 g/day on higher-quality dried grass alone can similarly be expected to gain 1.2 kg/day on a diet of dried grass plus concentrates.

Hay made with preservative

An alternative to the barn-drying of moist hay is to bale the crop with the addition of a preservative (usually based on propionic acid, see Ch. 3 p. 44). Relatively little research has been conducted to compare the two techniques, but the data in Table 5.2 show that, with uniform application of preservative to give effective prevention of moulding during storage, the performance of beef cattle (initially 12 months of age) is likely to be very similar to that of animals given barn-dried hay.

Dried legumes

Legume crops such as lucerne (alfalfa) are well suited to artificial dehydration, but less so to haymaking because of the risk of loss of leaf during field-drying.

Table 5.2 Performance of beef cattle given barn-dried or chemically-preserved hay

	Barn dried	Treated with preservative*
Intake of DM (kg/day)	6.94	6.89
LW gain (kg/day)	0.87	0.90

* Ammonium bispropanoate, added at 1.5% of hay fresh weight
Source: data provided by Dr M. V. Tas

Table 5.3 The relative intake of legume and grass forages at 65% D-value

	Intake
Ryegrass	100
Lucerne	105
Red clover	115
Sainfoin	130

Apart from the advantage the legume has over grass in being able to fix N from the atmosphere, the different structure of the plant renders it more rapidly digested in the rumen. Thus, at the same digestibility, the intake of legume forages is usually higher than that of ryegrass (Table 5.3).

Studies of the performance of beef cattle given dried lucerne pellets as a replacement for cereal concentrates in cereal beef production showed that, despite an improvement in energy intake, weight gain and efficiency of feed use were reduced (Table 5.4). However, gains in excess of 0.9 kg/day were

Table 5.4 Performance of Friesian beef cattle given dried lucerne or cereal concentrates in a cereal beef system

	100 % dried lucerne pellets	50% lucerne/50% cereal concentrates	100 % cereal concentrates*
Intake of metabolizable energy (ME) (MJ per 100 kg LW per day)	37.0	34.0	30.6
Live-weight gain (LWG) (kg/day)	0.92	0.99	1.10
Efficiency of feed use (kg LWG per 100 MJ ME)	2.5	2.9	3.6

* 85% rolled barley, 15% protein/mineral balancer

Source: from Bastiman, B. *et al.* (1982) *Experimental Husbandry, No. 38*, p. 99

recorded from dried lucerne offered *ad libitum* as the sole feed, which contained only 9.6 MJ ME per kg DM. In this trial, intake of DM for the lucerne diet was high, averaging 36.2 g/kg live weight. Further, the drying of this relatively high-protein material at high temperature would have been reflected in a reduction in the degradability of the protein fraction, thereby enhancing the supply of undergraded dietary protein to the animal.

Dried forage maize

In areas suitable for the growth of legumes for dehydration it is also common to grow forage maize for dehydration, since the harvest of maize comes at the end of the growing season and can usefully increase the annual output of the dryer. Not surprisingly, dried whole-crop maize with its relatively high (50%) content of ear in the total DM is a high-energy, but relatively low-protein product. The major problem with its production is that the plant is not uniform in terms of DM content at harvest. The ear is usually much drier than the stover. Thus care must be taken to avoid over-drying the ear and thereby damaging the grain.

Experiments in France and Belguim have demonstrated that bulls may be finished in a feedlot beef system on dried whole-crop maize, and achieve rates of live-weight gain and levels of efficiency of feed use similar to those obtained from comparable animals given maize silage (Table 5.5).

Although the intake of pellets was 22 per cent greater than that of silage, weight gain was also higher and efficiency of feed use only slightly lower than that of bulls given the maize silage diet. The work in France by the Institute Technique des Cereales et Fourrages in 1976 (Table 5.6) has examined the effect of feeding dried whole-crop maize in different physical forms to bulls in feedlot beef production. In this trial, maize comprised 80 per cent of the total diet DM, and supplements of protein, minerals and vitamins were given together with straw (at 6% of the diet) to provide essential long fibre in the diet. Levels of intake and weight gains were similar between wafers, cobs and pellets.

Compared with maize silage, losses during dehydration may be expected to be relatively low (5% for the dried crop compared with 22% for the ensiled crop; Table 5.7). Thus, despite a slightly lower efficiency of feed use, output of live-weight gain per hectare from dried maize was some 12 per cent higher than for silage, at a similar input of concentrate supplement per hectare (Table 5.7).

Table 5.5   Dried whole-crop maize compared with maize silage in feedlot beef production using bulls

	Maize silage	Dried whole-crop maize pellets
Initial LW (kg)	278	278
LW at slaughter (kg)	565	575
Intake of DM (g/kg LW)		
Concentrate	8.1	8.1
Maize silage	9.6	–
Dried maize	–	11.7
Total	17.7	19.8
LWG (kg/day)	1.17	1.23
Efficiency of feed use		
(kg LWG per 100 kg DM intake)	15.8	14.5

Source: after Boucqué, C. V. *et al.* (1976) *Animal Feed Science and Technology* **1**:347

Table 5.6   Physical form of dried whole-crop maize and its influence on feed value for young bulls in feedlot beef production

	Physical form of dried whole-crop maize		
	Wafers*	Cobs[†]	Pellets[‡]
Initial LW (kg)	154	154	155
Final LW (kg)	573	572	591
Intake of DM (g/kg LW)			
Dried maize	19.9	19.8	19.6
Supplement	5.0	4.9	4.9
Total	24.8	24.7	24.5
LWG (kg/day)	1.24	1.19	1.26
Efficiency of feed use			
(kg LWG per 100 kg DM intake)	13.8	13.3	13.8

* Wafers: unprocessed dried material, pressed into wafers of relatively large diameter
[†] Cobs: unmilled dried material, extruded through a die-press
[‡] Pellets: milled dried material, extruded through a die-press

Table 5.7   *Production of beef per hectare of land: dried whole-crop maize compared with maize silage*

	Maize silage	Dried whole-crop maize
Yield of DM (kg/ha)	12 250	12 250
Useful feed (%)	78	95
Yield of useful feed DM (kg/ha)	9 518	11 637
Efficiency of feed use (from Table 5.5)	15.8	14.5
LWG per ha (kg)	1 504	1 687

Source: from Boucqué, C. V. *et al.* (1976) *Animal Feed Science and Technology* **1**:347

Systems of production from dried forages

From a knowledge of the levels of intake and performance that can be achieved by beef cattle given dried forages of known quality *ad libitum*, target rates of weight gain and feed budgets can be set for systems of production.

Physical targets for the performance and feed use for beef cattle given dried forages are in given Table 5.8. Three levels of forage quality are specified and here the feed budgets reflect the need to supply a defined amount of energy to the animal to meet the target live-weight gain for the particular system. Requirements for concentrate increase substantially with decreasing content of ME in the dried forage; this reflects not only the lower content of energy but also the lower level of intake (Table 4.7). Requirements for dried forage are lower at the lower contents of ME. Combined with an increased yield per hectare of the lower-quality material, the use of such material would be appropriate for farms where land for forage conservation is limited in relation to the number of animals on the holding.

There have been few studies of beef systems in which cereal concentrates have been completely replaced by dried forages. In one trial with bulls, dried grass was compared with barley in an 18-month (grass/cereal) system. The results for performance and feed consumption by the cattle are in Table 5.9: they show little difference between dried forage (which was of high quality; 69% D-value) and cereal-grain as supplements to grazed pasture and to silage, although the total amount of supplement given to the cattle in this trial was quite high.

Table 5.8 Beef from dried forage: targets for performance and target feed budgets

	System								
	Grass/cereal beef*			Grass beef†			Feedlot beef		
Age at slaughter (months)	18			22			16		
Initial LW (kg)	50			50			300		
LW at slaughter (kg)	500			520			450		
LWG (kg/day)	0.8			0.7			0.8		
Feed budgets									
Quality of dried forage (MJ ME per kg DM)	10.0	9.0	8.0	10.0	9.0	8.0	10.0	9.0	8.0
Dried forage DM (t per head)	1.4	1.25	1.0	1.4	1.25	1.1	1.25	1.2	1.1
Concentrate DM (t per head)	0.4	0.6	0.9	0.2	0.3	0.5	0.2	0.3	0.4
Grazed grass DM (t per head)	1.5	1.5	1.5	3.5	3.5	3.5	—	—	—

* Grazed at pasture, 6 to 12 months of age (180–330 kg LW)
† Grazed at pasture, 4 to 10 months of age (125–270 kg LW) and from 16 to 22 months of age (400–520 kg LW)
‡ Offered to the cattle

An assessment of the effect of hay quality on feed budgets in grass/cereal beef production is given in Table 5.10. Ryegrass, cut at 59 per cent D-value (8.8 MJ ME per kg DM) is conserved either by field-drying or barn-drying, with or without exposure to rain damage. The longer period of exposure to the weather in the case of field-dried hay is reflected in a lower digestibility compared with barn-dried material, particularly when rain occurs. This, combined with a reduction in the potential intake of the hay, is associated with an increase in the requirement for concentrates to meet the specified rates of live-weight gain during the winter periods (0.7 and 0.8 in the first and second winters, respectively). The increase in concentrate requirements associated with rain damage is 30 per cent in the case of the field-dried hay, and 20 per cent in the case of barn-dried material.

In practice, the choice as to whether hay is field- or barn-dried is rarely made on an individual farm; those farmers in possession of barn-drying equipment generally take advantage of it to reduce the risk of rain damage. The additional cost of so doing is balanced by the reduced cost of concentrates, so that margins per

Figure 5.2   Large bales can speed up haymaking, but they are difficult to barn-dry. An alternative method of conserving moist hay made in large packages is to seal it in polyethylene

Table 5.9   *Grass/cereal beef: comparison of dried grass with barley as supplements to silage and grazed pasture; levels of performance and feed use by Friesian bulls*

	Supplement	
	Barley	Dried grass
Initial LW (kg)	48	48
LW at slaughter (kg)	552	536
Age at slaughter (months)	18	18
LWG (kg/day)	0.93	0.89
Total feed use (kg DM)		
Silage	971	976
Supplement	1 032	1 001

Source: after Tayler J. C. *et al.* (1974) *Technical Report Grassland Research Institute, Hurley, No. 13*

Table 5.10   Grass/cereal beef*: effect of hay quality on winter feed budgets

	Barn-dried hay		Field-dried hay	
Rain damage	−	+	−	+
Digestibility of hay (D-value, %) (59% D-value at cutting)	55	47	57	54
Feed budget (t DM per head)				
Hay	1.01	0.86	1.08	0.98
Concentrates	0.56	0.73	0.49	0.59

* Hereford × Friesian steers
Source: data from Dr S. Marsden

head are likely to be similar. But because the intake of hay is greater for barn-dried than for field-dried material, requirement exceeds the benefit to yield associated with lower field losses and margins per hectare of land are likely to be somewhat lower when hay is barn-dried.

Chapter 6 **Beef from silage**

The majority of the larger beef producers use silage as their basic forage rather than hay, for reasons that are as much associated with ease of management of the enterprise as with the nutrition of the beef animal. Silage is better suited than hay to mechanization and to the speedy harvest of large amounts of forage. Self-propelled forage harvesters are now available that can harvest forage crops of grass or forage maize at the rate of up to 30 tonnes fresh crop per hour. Removal of silage from store can be automated and, with mixer wagons, complete feeds based on silage can be distributed mechanically to very large numbers of animals.

But the extent to which farmers rely on silage and grazed grass as sources of nutrients to the beef animal is quite low. An analysis of the contribution made to the winter requirement for metabolizable energy (ME) by beef cattle showed that, at the levels of concentrate used on progressive enterprises where feed use and animal performance were recorded, less than 60 per cent of the ME required to support the recorded levels of live-weight gain came from silage (Table 6.1).

Thus, although there may be sound economic and managerial reasons for wishing to produce beef from silage, problems may be encountered in achieving high rates of animal growth from diets in which silage is the major source of nutrients.

Quality of silage and growth of beef cattle
Pattern of fermentation

An important influence on the value of silage to the beef animal is the quality of preservation. The occurrence of extensive secondary (butyric) fermentation in the silo not only leads to increased loss of nutrients, but also to reduced consumption. These effects can be seen particularly in young animals, but they also occur in older cattle. Reductions in dry matter (DM) intake of 30 per cent are not uncommon and carcass weight gains can be halved (also see Table 4.6).

Table 6.1 Contribution made by silage to the winter requirement for metabolizable energy in systems of beef production in the UK

	System	
	Feedlot beef*	Grass/cereal Beef
Live-weight gain (LWG) (kg/day)	0.8	0.8
Total requirement for metabolizable energy (ME) (MJ)[†] Of which	11 160	27 445
Silage	6 445	7 680
Concentrates	4 715	10 855
Grazed grass	–	8 910
ME from silage (% of silage ME + concentrate ME)	58	41

* 180 days on feed, from 300 to 450 kg live weight
[†] from Table 2.9

Degradation of nitrogen to ammonia, mentioned in Chapter 5 as a factor that leads farmers to give hay rather than silage to calves, is reflected in a reduced supply of amino acids for lean tissue growth, so that even in the young animal a relatively high proportion of the weight gain may be in the form of fatty tissue compared with dried forage diets (Table 6.2). Thus, at equal energy intake above that required for maintenance, the weight gain achieved from the diet based on silage would be 40 per cent lower than that on the dried grass diet. Essentially similar results were obtained in the trial illustrated in Table 4.1.

Table 6.2 The energy value of the weight gains of young beef cattle given diets based on silage or dried grass

	Diet	
	Dried grass	Silage
Empty body weight gain (kg/day)	0.5	0.5
Energy value of weight gain (MJ/kg)	9	15
Tissue deposition (g/day)		
Protein	107	76
Fatty tissue	53	142

Source: from Lunsdale, C.R. (1976) *Ph.D. Thesis, University of Reading*

Dry-matter content   Wet, direct-cut crops are at greater risk than drier, wilted ones with respect to secondary fermentation. Thus comparisons show better performance by growing cattle from wilted silage than from direct-cut material. The advantage from wilting is usually 100–150 g/day extra live-weight gain. But the disadvantage to direct-cut silage can be offset by use of concentrate supplements, or by use of an effective additive to ensure good preservation.

In practice, therefore, an additive is more likely to be used on wetter than on drier crops. Much of the benefit of wilting to beef cattle growth can be attributed to an improvement in quality of preservation, rather than to the increase in DM content *per se*. There is now clear evidence that, when the direct-cut crop is well preserved, there is no benefit to be obtained from wilting, in terms of either the performance of beef cattle or the efficiency with which they convert silage to carcass gain. Although intake of wilted material can be 20 per cent greater than for direct-cut silage, digestibility is usually slightly lower. Carcass gain is not necessarily improved by wilting and efficiency of feed use can consequently be lower.

The wilting of silage for beef cattle can therefore only be justified in terms of a greater speed of harvest, a reduction in the problem of silage effluent and a lower cost of additive.

Evidence is now accumulating to indicate that there is a reduced supply of amino acids from wilted compared with well preserved direct-cut silage; this, together with a lower digestibility, may explain the lack of response in carcass gain to wilting.

Digestibility   Despite the fact that pattern of fermentation can often override digestibility in its effect on intake and performance, an analysis of trials in which different silages were given as the sole feed to beef cattle revealed the relationship shown in Fig. 6.1. While intake is related to digestibility, pH and the content of DM in the silage (see Ch. 4), live-weight gain is apparently affected only by digestibility. The response in gain was 35 grams for each percentage unit increase in digestibility of DM.

The potential live-weight gain from silage given as the sole feed is 0.8 to 1.1 kg/day, depending upon the growth potential (breed and sex) of the animal. But to achieve these levels of performance it is necessary not only to harvest grass at an early stage of growth (in excess of 70% D-value), but also to preserve it well in the silo. in the silo.

Figure 6.1   Response in live-weight gain by 400 kg beef cattle to improving silage quality. LWG = 0.0346 DMD − 1.763

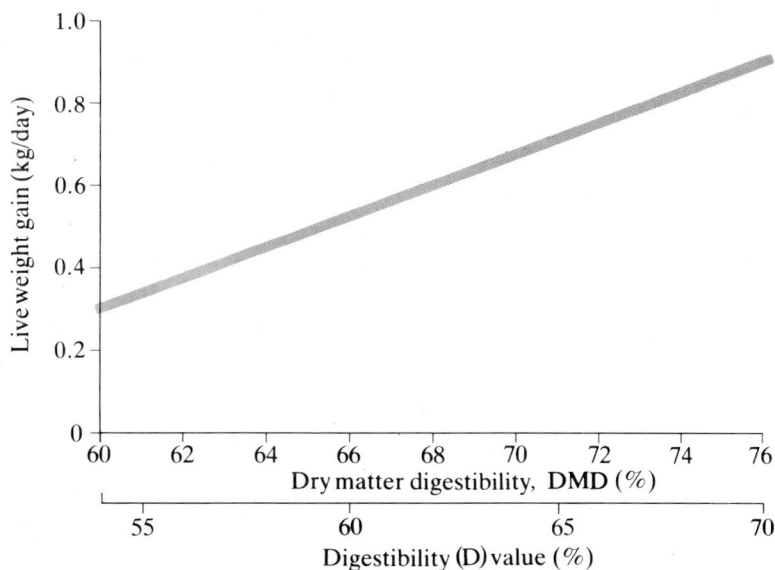

Additives   The beneficial effect of additives (e.g. formic acid, formalin and acid, and molasses) on the fermentation of silages has already been described (Ch. 3); in particular, those that not only acidify, but which also discriminate against the clostridial bacteria, are likely to be most effective in improving intake and weight gains by beef cattle.

In recent years, additives based on mixtures of formalin (35% formaldehyde) and acid have been introduced. The objective is

The growth of beef cattle in 26 paired comparisons of untreated silage, with the same crop treated with formic acid at harvest (mean 2.6 litres/t; range 1.0–4.81 litres/t), is shown in Fig. 6.2. In all but two trials, when there was no response to the additive, performance was improved, in some cases by a large amount. The average improvement in rate of live-weight gain was 0.2 kg per head per day. Surprisingly, the improvement was consistent across the whole range of weight gains. This evidence confirms the effectiveness of formic acid as an additive in improving the quality of preservation and reducing loss of energy during storage.

Figure 6.2   Live-weight gains (kg/day) by beef cattle given untreated or formic acid-treated silage. The $y=x$ line shows the position of equal response. Almost all the values lie above the line, and it may be concluded that silage treated with formic acid is superior to untreated silage at both low and high live-weight gains

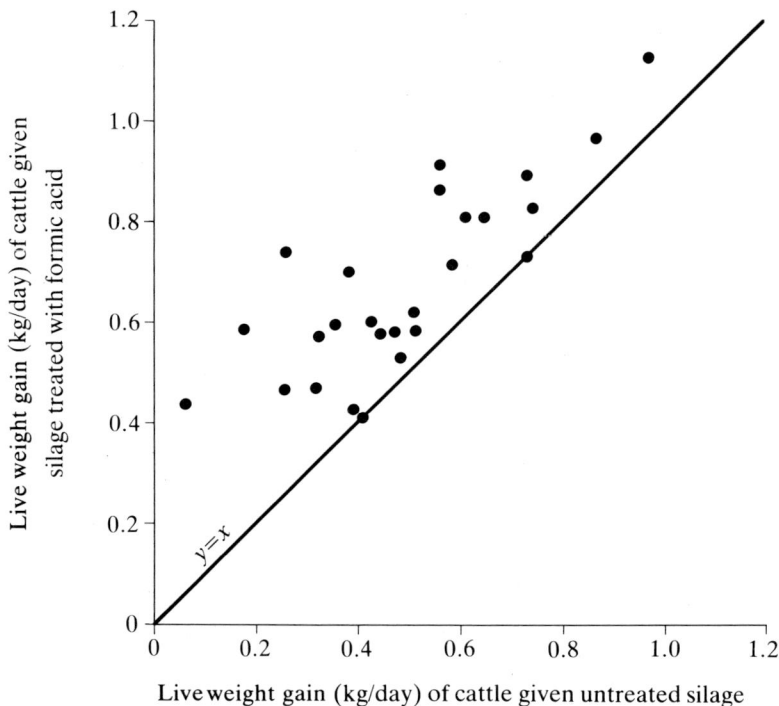

Live weight gain (kg/day) of cattle given untreated silage

to reduce the extent to which protein is degraded, so that the animal receives an increased supply of undegraded dietary protein (UDP). The date in Fig. 6.3 indicate little overall response in beef growth to formalin/acid mixtures over the rate of growth of cattle given silage made from the same initial crop but treated with formic acid. Some of the negative responses may have been due to over-protection of protein as a result of the use of relatively high levels of formalin in relation to the content of protein in the crop.

Alternatively, the animals used to evaluate the additives may have been insensitive to an increase in the supply of undegraded protein; they may have been too high in weight to require UDP

Figure 6.3   Live-weight gains (kg/day) by beef cattle given formic acid or formalin
plus acid-treated silage

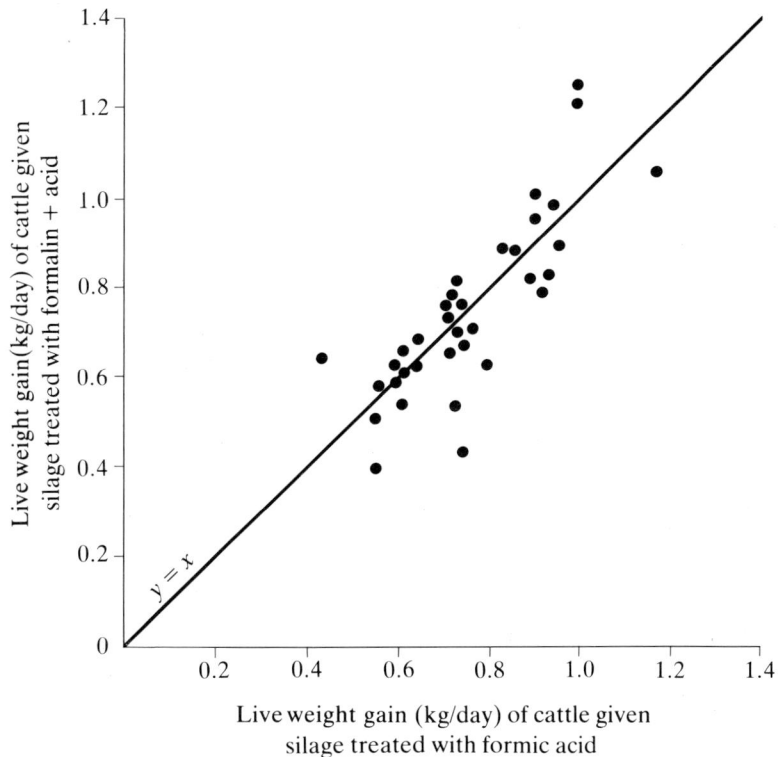

of feed origin, in addition to the protein supplied from microbial
synthesis in the rumen.

In a trial with cattle of relatively high potential for lean tissue
growth, the use of a mixture of formalin and formic acid was
superior to formic acid added alone to a ryegrass crop (Table
6.3). Intake of ME and live-weight gain were 10 and 18 per cent
higher, respectively, in the case of the silage treated with the
mixture of formalin and acid. The carcasses from the animals
given the silage made with the mixed additive were leaner than
those from the animals given the silage made with formic acid
alone, reflecting not only the higher energy intake, but also a
superior retention of dietary nitrogen.

Table 6.3   *Response of Charolais-cross cattle to formalin and formic acid-treated silage*

	Formic Acid*	Formalin and formic acid[†]
Dry matter %	27.2	28.0
pH value	4.1	4.1
ME content of silage (MJ/kg dry matter)	10.3	11.1
Intake of ME (MJ per 100 kg live-weight per day)[‡]	18.4	20.2
LWG (kg/day)	0.83	0.98
Fat % of rib sample joint	31	29
Lean : fat ratio	1.7	1.9

* 3.7 kg/t fresh crop
[†] 3.4 formic acid per t + 1.4 kg formalin per t
[‡] 1.5 kg barley supplement per day
Source: adapted from the work of Dr C. E. Hinks and co-workers at the Edinburgh School of Agriculture

Supplementary energy

The response in beef cattle growth to supplementation with concentrates is generally considered to be mainly a reflexion of silage quality. Other factors that might influence the response include the level of supplementation and the nature of the supplement itself.

The response to additional concentrate depends to a large extent on the level of live-weight gain supported by the unsupplemented diet (Fig. 6.4). Thus at high live-weight gain (above 0.75 kg/day), the response in weight gain to concentrates is negligible, whilst with poorer quality silage, capable of supporting only 0.20 kg/day weight gain, the response was high, at 0.20 kg/kg concentrate.

A review of 13 trials in France with cattle of relatively high growth potential showed that the response to 1 kilogram extra concentrate per head per day, above a 'control' level of 1–2 kilograms (0.3–0.6% of live weight), was 0.10 kg extra daily live-weight gain. This response occurred irrespective of the content of DM in the silage, use of additive, type of animal or the level of live-weight gain supported by the control diet (which was high, ranging from 0.84 to 1.48 kg/day). The additional concentrate, however, increased total DM intake (by 0.34 kg)

Figure 6.4   Effect of the level of live-weight gain achieved from silage alone on the response in gain to additional concentrates. The vertical axis shows the live-weight gain achievable from silage alone and is a measure of silage quality. The horizontal axis shows the response to added concentrates in terms of the live-weight gain achieved per kilogram of extra concentrates given. Thus, for good silages able to support more than 0.75 kg daily live-weight gain alone, there is little response to concentrates. Whereas, for silages only able to support 0.5 kilogram daily live-weight gain or less, there can be 0.1–0.2 kilogram extra gain for each kilogram of concentrates

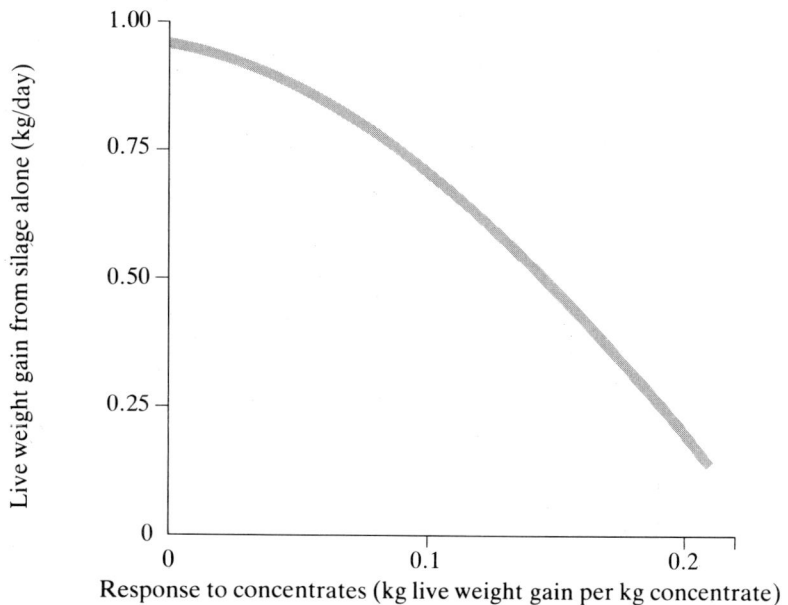

and decreased silage intake (substitution value 0.66), and there was no improvement in efficiency of feed use. The additional ME supplied by the extra cereal supplement tended to be reflected in increased carcass fatness. In these trials, all silages were of high digestibility and were well preserved.

Much lower responses have been recorded at the Grassland Research Institute with maize silage (Table 6.4). This is not surprising because, once the deficiencies in protein and minerals are rectified, high rates of weight gain are achievable. Thus bulls of high growth potential gave weight gains of almost 1.3 kg/day

Table 6.4   *Effect of level of maize grain supplement on live-weight gain and efficiency of feed use by bulls given maize silage*

	Maize grain		
	0	1	2
Intake of DM (kg/day) of which	6.6	6.7	7.2
Silage	5.4	4.6	4.2
Protein concentrate	1.2	1.2	1.2
Maize grain	—	0.9	1.8
LWG (kg/day)	1.28	1.30	1.34
Efficiency of feed use (kg LWG per 100 kg DM intake)	19.2	19.4	18.5

on a diet of maize silage and protein concentrate. Additional energy in the form of maize grain gave a response of only 0.025 kg extra live-weight gain per kilogram supplementary grain, with no improvement in efficiency of feed use. However, with cattle of this type it is advisable to include some supplementary energy in the diet during the final finishing period, to ensure that the carcass has an adequate level of fatness for the meat trade.

Supplementary protein   In view of the fact that a high proportion of the protein in silage is degraded during the ensiling process (Fig. 4.1), the inclusion of supplementary protein in diets based on grass silage might be expected to lead to responses in performance, particularly if the animal has a relatively high requirement for tissue protein for growth of muscle. Thus there has been a considerable amount of interest in the extent to which beef cattle might respond to supplementary protein. An analysis of 27 comparisons, of which 16 involved fish meal and 11 soya bean meal, revealed a positive effect on live-weight gain in all but five (Fig. 6.5). There is some debate as to whether the response is to protein or to energy. In most cases, the inclusion of supplementary protein in the diet did not depress the intake of silage as much as would be expected; in some cases, intake of silage was actually increased following the addition of protein to the diet. Thus intake of energy was

Figure 6.5   Response by grazing beef cattle to the supplementation of grass silage diets with extra protein

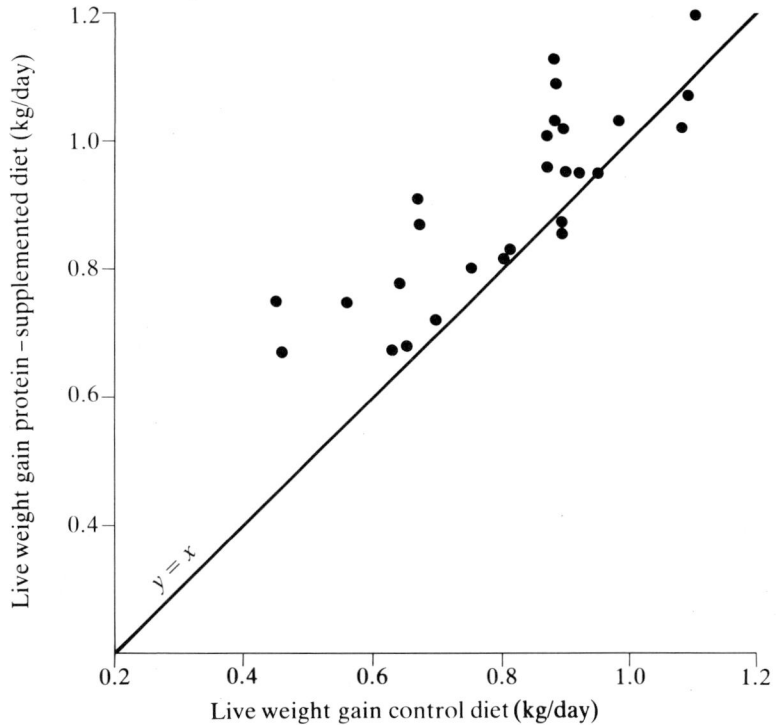

Figure 6.6
Supplementation of grass silage with fish meal: response of calves of less than 20 weeks of age to increased intake of undergraded dietary protein

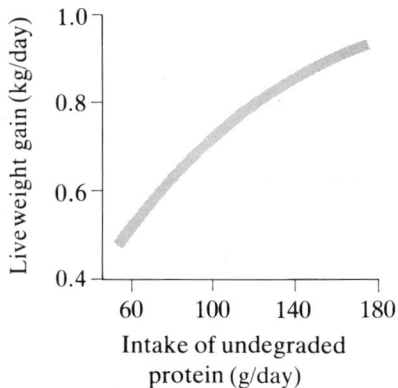

increased as well as that of protein.

With young calves, of less than 150 kilograms live weight, diets of grass silage supplemented with different amounts of fish meal appear to support levels of live-weight gain that closely reflect the intake of UDP (Fig. 6.6). Older calves, in excess of 150 kilograms live weight, are likely to give levels of weight gain that correspond to the intake of ME from the diet rather than to the intake of UDP (Fig. 6.7). But at high levels of supplementation with fish meal (more than 0.5g/kg live weight), live-weight gain may exceed that that would be expected from the intake of ME.

A further factor to be borne in mind is that of the duration of the response to supplementary protein. In some trials, the response to protein during the winter period has not been maintained subsequently whilst at pasture. Thus a cheaper strategy for systems of production that involve a grazing season may be to

Figure 6.7   Relationships between live-weight gain and intake of metabolizable energy by calves of 150 kilograms. Whereas Fig. 6.6 showed the response of younger calves to undergraded dietary protein, this Fig. shows that, for older calves, the major factor affecting the rate of daily live-weight gain is the energy intake

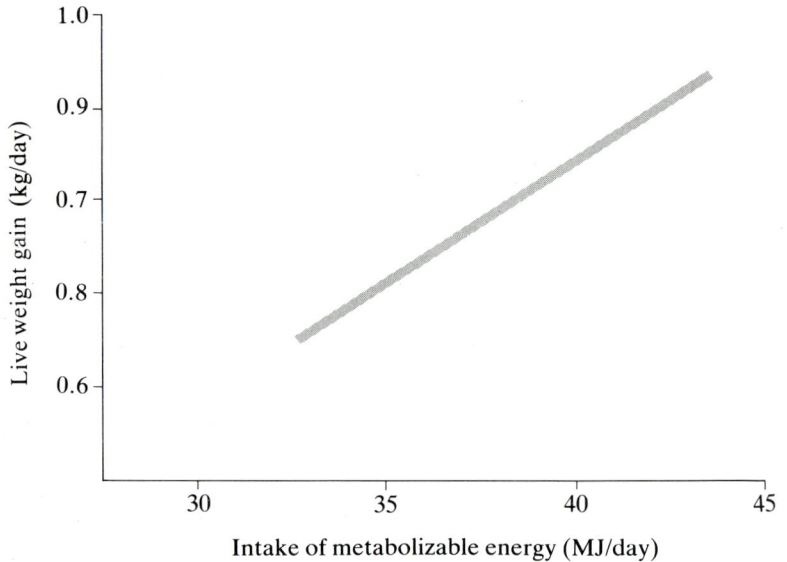

accept a lower rate of gain in winter and then to ensure adequate feed intake at pasture, so as to exploit compensatory growth as much as possible.

Systems of beef production from silage   Targets for performance are set in relation to breed, and the age and weight at which a particular breed reaches a suitable level of carcass fatness for the meat trade (see Fig. 2.2). In this section, the effect of variation in the quality of silage on feed requirements is examined for a range of systems. As with the production of beef from dried forage (Table 5.8), a decline in the energy value of silage is reflected in an increase in the amount of concentrate that is required. It should be stressed that the values for the feed budgets in Table 6.5 are for well preserved silage of high intake potential. The intention is to maximize the contribution of silage in the diet and thereby minimize the input of concentrates.

Table 6.5 *Beef from silage: targets for performance and target feed budgets. With grass silage the quantity of concentrate varies according to the quality of the silage, so that energy intake and target daily live-weight gains are achieved*

	System									
	Grass/cereal beef*			Feedlot beef*			Silage beef†			
							Grass			Maize
Age at slaughter‡	18			18			15			15
LWG (kg/day)‡	0.8			0.8			1.0			1.0
LW at slaughter (kg)	500			450			500			500
Feed budgets										
Quality of silage (MJ ME per kg DM)	10.0	9.5	9.0	10.0	9.5	9.0	10.0	9.5	9.0	10.7
Silage DM (t per head)	1.4	1.30	1.25	1.20	1.15	1.1	1.7	1.6	1.55	1.7
Concentrate DM (t per head)	0.5	0.6	0.8	0.2	0.3	0.4	0.6	0.7	0.8	0.5
Grazed grass (t per head)§	1.5	1.5	1.5	–	–	–	–	–	–	–

* Weaned suckled steers of breeds of medium mature size, finished on grass silage
† Bulls of breeds of medium mature size (see also Figs 6.8 and 6.11)
‡ See also Table 2.7
§ Offered to the cattle

Grass/cereal beef

With grass/cereal beef it is important that weight gain at pasture (from 6 to 12 months of age) is as high as possible. The targets of 0.8 kg/day for Friesian steers and 0.9 kg/day for Hereford × Friesian steers are critical. Gains not made in the grazing season are very difficult to make good in the following winter period. It is important, therefore, not to sacrifice performance at grass by setting aside too great an area for first-cut silage, in an attempt to make a large quantity of high-quality conserved forage from a limited total area of land.

The area of land to be allocated to grazing until silage regrowths are available depends on the quality of the land, assuming a total application of N fertilizer of 150 kg/ha. Below-average land will probably only support about 1 000 kg of cattle live weight per hectare at turnout, whereas average and above-average land should support 1 400 and 1 600 kg/ha, respectively. If the cattle weigh about 180 kilograms live weight at turnout, this means 5.5 head per hectare turned out on to below-average land and 9 head per hectare on to above-average land. To give an idea of areas per head, these would be about 0.2, 0.15 and 0.10 hectares per head for the three qualities of land respectively.

Similarly, the amount of land required to meet the needs of silage will differ according to the quality of the land. Allocations of land to first-cut silage might be about 0.3 hectare per head for below-average land, 0.25 for average land and 0.20 for above-average land. With 100 kilograms N fertilizer applied per hectare, this should yield enough silage to provide 1.2 tonnes silage DM per head of cattle. If the need is for 1.5 tonnes silage DM, then land allocations should be raised to about 0.35, 0.30 and 0.25 hectares per head of cattle, respectively.

The plan is to graze the area set aside until regrowths from the first cut of silage are available and then to cut a proportion of the grazed area for silage in mid-season. The amount of land to be cut at this time depends on the total required and the yield of first-cut silage. The total grassland requirement, shown in Table 6.6, is the sum of the area for grazing and that for silage. After the first-cut silage has been taken, the area for grazing is expanded to include a greater proportion of the regrowth area as the season progresses until, in late-season, the whole area is grazed.

Table 6.6 *Total grassland requirements for grass/cereal beef production from silage. Land allocations are for the sum of the requirements for growing and the requirements for silage. Nitrogeneous fertilizer is assumed to have been applied at 150 kilograms nitrogen per hectare*

| Silage (t DM per head) | Quality of land | | |
	Below average	Average	Above average
	Total land required (ha per head)		
1.2	0.50	0.40	0.30
1.5	0.55	0.45	0.35
	Stocking rate (head per ha)		
1.2	2.0	2.5	3.3
1.5	1.8	2.2	2.9

A useful tactic to maintain live-weight gain from pasture in early season is to vary the area set aside for grazing by using a *buffer zone*. This zone, about one-third of the allocated grazing area, is separated from the grazing area by a movable electric fence. If grass is in short supply, the fence is moved to maintain a sward height of 50 millimetres over the whole grazing area. In seasons when growth of grass in early season is low, the buffer zone is completely grazed and the yield of silage is reduced. This shortfall may be rectified by increased fertilizer applications in mid-season, or by increased concentrate use in the following winter. Thus grazing is given priority in grass/cereal beef production, since failure to meet the target gain during the grazing season means that either the winter period is extended, or the amount of concentrates required for the finishing period is increased. In either case, margins are reduced.

Intensive beef from grass silage

It is one thing to say that the achievement of the target live-weight gain at pasture is vital to the success of grass/cereal beef; it is quite another matter to actually achieve the target. Indeed, the variability in rate of growth of grass, combined with the threat of parasitic worm infestation of the grazing animal, often results in poor gains, especially in mid- and late-season.

For farmers who have already invested in buildings and machinery for making silage and for feeding beef cattle on silage in the winter months, a logical consequence of poor performance at pasture is not to graze at all.

Data from nine experiments in which well preserved silage was given to Friesian or Friesian-cross cattle of known age and weight, either as the sole feed or with levels of supplement, are given in Fig. 6.8.

The results indicate that it should be entirely possible to achieve a target live-weight gain of 1.0 kg/day to give finished cattle at 15 months of age from a diet of high-quality grass silage (10.0 MJ ME per kg DM), with no more than 1.5 kilograms concentrate supplement per day. Indeed, the supplement may be offered at a flat rate for the entire feeding period (12 months) in a very simple feeding system. If this policy were to be adopted, the total requirement for concentrate would be about 0.6 tonnes per head (see Table 6.5).

There are a number of advantages to this intensive system. First, there are greater opportunities to control the nutrition, and

Figure 6.8   Intensive beef production from grass silage: results from experiments in the UK and France

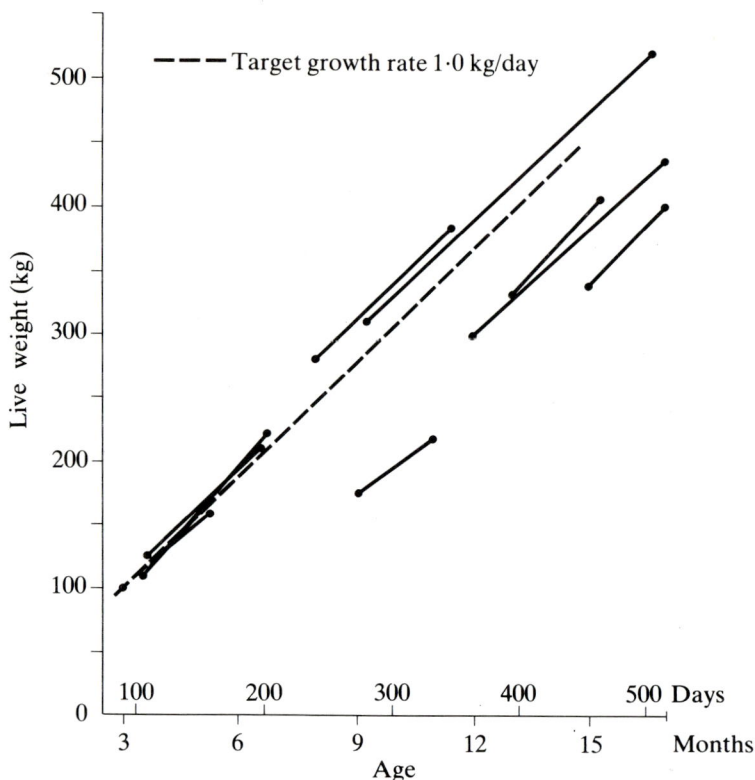

hence the rate of growth, of the animal than in the case of conventional grass/cereal beef. This allows greater budgetary precision and closer managerial control of the enterprise.

Secondly, the system can be operated continuously, with batches of calves being purchased and batches of finished cattle being sold each month of the year. This aids the pattern of cash flow (see Ch. 2).

Thirdly, the production of silage can be considered in a similar way to the production of an arable crop; varieties of grass can be chosen to suit a cutting strategy rather than having to be subjected to both cutting and grazing management.

Fourthly, there are opportunities to increase output per hectare of land by reducing wastage, which can often occur with grazing.

Figure 6.9   Intensive beef from grass silage. There is greater opportunity to control the animal's diet and its rate of growth than when it grazes at pasture

But there are disadvantages. The costs of machinery and labour are likely to be higher because, not only is it necessary to feed the cattle all the year, more manure must be moved out of buildings, which are more fully occupied than would otherwise be the case.

Tests of the system have been few, but trials at Rosemaund Experimental Husbandry Farm, reported to the British Society of Animal Production by Mr R. Hardy in 1981 (Table 6.7), have shown that the target rate of gain can be consistently maintained – albeit with a rather high level of concentrate given to bulls of relatively small mature size (Hereford × Friesian), which resulted in a low weight at slaughter. Nevertheless, at a stocking rate of eight cattle per hectare of grass, there is clearly scope for achieving relatively high margins from the system.

Table 6.7 *Intensive beef production from grass silage: results of a test of the system with Hereford × Friesian bulls*

Number of animals		19	
Initial LW (kg)		107	
Initial age (weeks)		10	
LWG			
10–42 weeks of age (107–327 kg LW)		0.98	
42–52 weeks of age (327–407 kg LW)		1.08	
Overall		1.01	
Feed intake (t DM)	Silage*		Concentrate
10–42 weeks	0.59		0.49
42–52 weeks	0.35		0.12
Total	0.94		0.61
Days on feed	296		
LW at slaughter (kg)	407		
Carcass weight (kg)	221		
Stocking rate (cattle per ha)	8.0		

* 10.2 MJ ME per kg DM

Maize silage

The maize crop is one of the most important arable crops in the world, with the capacity to yield over 10 tonnes DM per hectare of forage with a content of ME averaging 10.7 MJ/kg DM (see Ch. 3).

When harvested without field-wilting in late September or October, the crop contains adequate fermentable carbohydrate to ensure a lactic acid-dominant fermentation in the silo. Levels of DM intake by cattle are consequently relatively high. However, the content of crude protein is usually low (8–9% of the DM) and supplementation with additional protein is essential in the case of young beef cattle. Older cattle (in excess of 230 kg live weight) have shown satisfactory performance when given maize silage supplemented solely by non-protein N in the form of urea, with minerals and vitamins (Table 6.8).

This trial, and others in the UK, France and Italy, in which the amount of supplement has been restricted to a small quantity of a protein-rich concentrate or urea, have demonstrated that a target rate of live-weight gain of 1.0 kg/day is achievable from 100

Table 6.8   *Performance of Friesian steers at different ages given maize silage ad libitum with or without urea as the sole supplement**

	Age (months)					
	4.5	7.5	10.5			
LW (kg)	144	239	310			
Urea	−	+	−	+	−	+
Intake of DM (g/kg LW)	21.6	22.3	20.9	23.7	21.2	21.4
LWG (kg/day)	0.39	0.56	0.59	1.03	0.95	1.06

* A mineral/vitamin block was also on offer.
Source: after Thomas, C. *et. al.* (1975) *Journal of Agricultural Science (Cambridge)* **84**: 353

Figure 6.10   Intensive beef production from maize silage: results from experiments in the UK and France. From Sheldrick, R. D. and Wilkinson, J. M. (1980) *ADAS Quarterly Review, No. 37*, p. 95

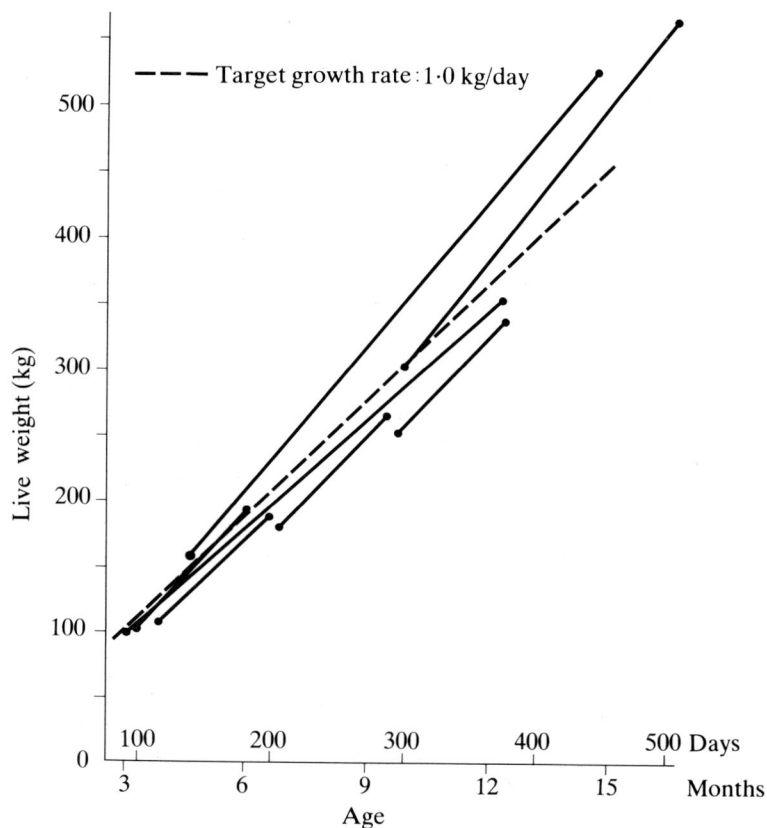

kilograms live weight (3 months of age) to slaughter at 450 or 500 kilograms (depending on breed and sex of animal) at 15 months of age (Fig. 6.10).

Further, there is evidence that, in countries where a choice can be made as to the proportion of the maize crop that can be harvested as the whole crop for silage (with the remainder being harvested as grain), the trend has been to reduce the proportion of grain in the total diet. The consequence of this has been an increase in stocking rate, a reduction in average daily gain and a reduction in carcass fatness. The effect on stocking rate and on weight gain of increasing the proportion of silage in the diet of finishing bulls is illustrated in Table 6.9.

Table 6.9   *Performance of bulls* given rations varying in the proportion of maize silage and maize grain in Italy*

	Total area harvested as silage (%)			
	100	84	56	38
Intake of DM (kg/day)	7.15	6.84	7.99	8.45
Of which (%)				
Maize silage	88	79	63	50
Maize grain	0	9	26	40
Protein supplement	12	12	11	10
LWG (kg/day)	1.0	1.05	1.1	1.2
Days on feed	220	210	200	183
Stocking rate (cattle per ha)	6.4	5.4	4.2	3.5
LWG per ha (kg)	1 408	1 191	924	769

* From 230 to 450 kg live weight
Source: after Kilkenny J. B. (1978) In *Forage Maize: Production and Utilisation* Agricultural Research Council, London

Tests of the intensive system for the production of beef from maize silage at three centres in the UK (Table 6.10) have shown that the targets in Table 6.5 are quite feasible. The feeding regime was simplified at centre C to the extent that a 'flat rate' of 1.5 kilograms supplement per day was given throughout the trial. The crude protein content of the supplement was 35 per cent of the DM. After the cattle reached 6 months of age, the cost of the diet was reduced by the inclusion of urea rather than vegetable protein in the supplement.

Results of the tests of the system (Table 6.10) show that the total quantity of maize silage required for the system, including losses during conservation, is about 1.7 tonnes DM per head. At a yield of 10 tonnes DM per hectare, this gives a 'stocking rate' of about six cattle per hectare.

Table 6.10 Intensive beef production from maize silage: results of tests of the system with Friesian cattle

Cattle	Centre		
	A Bulls	B Steers	C Bulls
Initial LW (kg)	100	111	92
Days on feed	352	304	367
Age at slaughter (months)	15	14	15
LW at slaughter (kg)	452	446	496
Daily LWG (kg)	1.0	1.1	1.1
Total concentrate DM (t per head)	0.53	0.53	0.62

Source: after Sheldrick R. D. and Wilkinson J. M. (1980) *ADAS Quarterly Review, No. 37*, p. 95

Chapter 7 Beef from conserved by-products

By-products, especially those from arable crops, have traditionally played an important role in the winter feeding of suckler cows and store cattle. Since this book is about the production of beef from conserved feeds, the emphasis here is on those by-products that are conserved prior to being used, either by drying or by ensiling. Not surprisingly, the crop by-products dominate and comprise principally straw. Recently, however, interest has been shown in the possibility that certain animal by-products, such as manure or digesta from slaughtered animals, may be recycled by conservation as silage in mixture with forages of other by-products.

Value of conserved by-products as feeds

The content of energy and nitrogen in some of the more common by-products is shown in Fig. 7.1. The range is enormous, from very low values for both energy and N in manure and straw, to very high values for N in blood meal and hydrolysed feather meal, and to very high values for energy in animal fat. these major compositional differences have important consequences in diet formulation, as illustrated in Fig. 7.2.

By-products and wastes may be classified into four categories:
A – *Low content of metabolisable energy (ME), low content of nitrogen.*
Products in this category comprise cattle faeces, straws, dried grape pulp and dried coffee residues. When the content of N is lower than that required by the rumen microbial population, the rate of fermentation can be reduced to the extent that these feeds are of very little value to the animal. Addition of degradable N

Figure 7.1   The content of metabolizable energy and nitrogen in some commonly used by-products

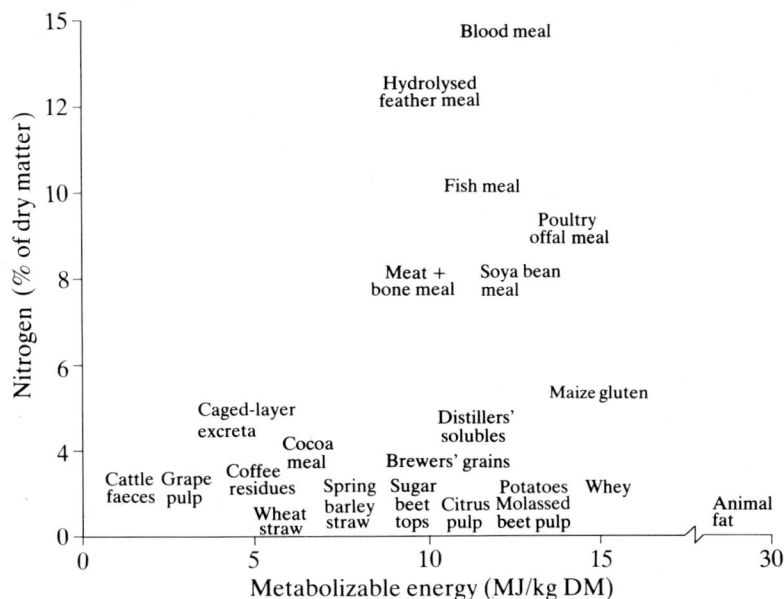

to the by-product or waste will assist in improving the value of the feed.

B – *Low content of ME, high content of N*
Few by-products come into this category, although caged layer excreta from hens and cocoa meal may be considered appropriate to this group (see Fig. 7.1). They contain an excessive amount of N in relation to their available energy and, as such, are complementary to the products in group A.

C – *High content of ME, low content of N*
Apart from animal fat, which has an exceptionally high energy value, the most common by-products in this group are those from the sugar industry, from the citrus fruit industry and from the milk industry (whey). When used in diets for cattle, they require additional N to satisfy the requirement of the microbial population. The rate of their digestion is usually very rapid and therefore the source of supplementary N must also be in a rapidly degradable form, such as urea.

Figure 7.2   Effect of the metabolizable energy (ME) content of feeds on the requirement for nitrogen by the rumen microbial population. The letter A denotes feeds of low ME and low N; B denotes low ME, high N; C denotes high ME, low N; and D denotes high ME, high N. The solid line gives the N needed by rumen microbes at various contents of ME in the feed. From Ørskov. E. R. (1980). In *BSAP Occasional Publication, No. 3*

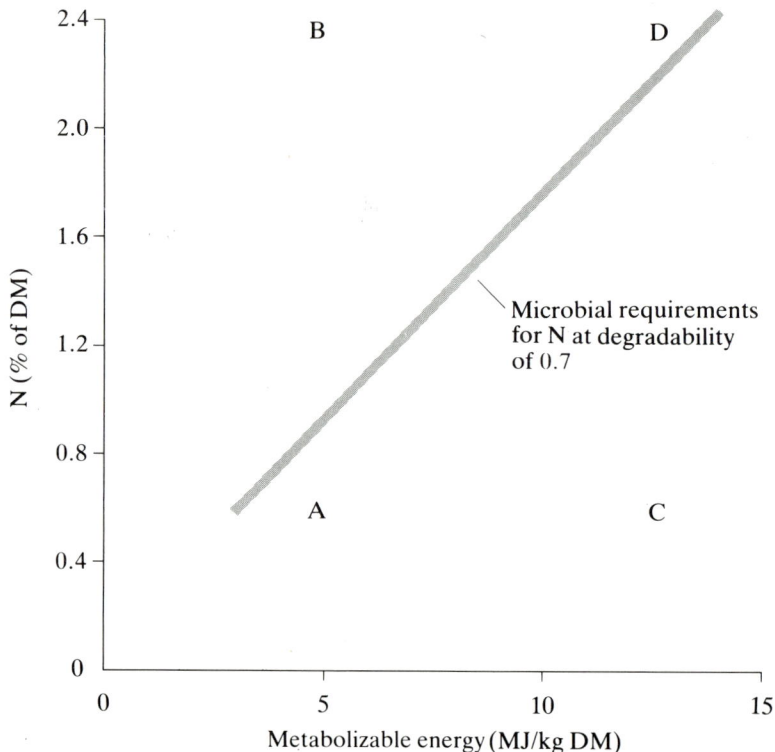

D – *High content of ME, high content of N*
Feeds in this category usually present few problems, other than of treatment for conservation. The N in blood meal, fish meal and poultry offal meal is, despite its high concentration, relatively undegradable in the rumen as a result of the drying process (and, in the case of fish meal, treatment with formalin). The protein in feather meal is indigestible unless the material is hydrolysed prior to use.

In general, the products in groups A and B are well suited for use in beef production, especially in diets for over-wintering of suckler cows and store cattle. They are particularly well suited as 'extenders' of conserved grass, such as low-quality hay, when supplies are limited. The products in groups C and D are better suited to those systems in which rapid weight gains are required; in particular, those in group C can usefully complement maize silage and those in group D high-quality grass silage.

In assessing the feasibility of including conserved by-products and wastes in diets for beef cattle, it is important to obtain expert advice as to the most appropriate level of inclusion, taking account of possible hazards (e.g. adverse effects on digestion and toxic factors).

Conserving by-products

Straw

Straw is most commonly conserved by baling after a short period of drying in the field. The bales may be small or large, of rectangular shape or round. Large bales are becoming increasingly

Figure 7.3   Harvesting straw directly from the swath by metered-chop forage harvester

popular because of their suitability to rapid, mechanized packaging, transportation and storage. Specialized containers may be used to hold the bales when they are used as feed.

Alternatively, straw can be harvested by forage harvester in similar manner to grass (Fig. 7.3). This technique is particularly useful if the straw is ultimately to be used in mixture with other feeds and dispensed to the cattle through a complete-diet mixer-wagon or a forage box. Calculations show that, even with a relatively small tractor, the number of man-hours per tonne of straw put into store is lower than that required to produce small rectangular bales, and to move them into store with a bale accumulator and conventional stacking (Table 7.1).

Table 7.1   *Forage-harvesting of straw compared with baling*

	Forage harvester*		Baler[†]
	Self propelled	Trailed	
Yield of straw (t dry matter (DM) per ha)	1.8	2.3	–
Rate of work			
Fill trailer/make bales (man-min/t DM)	19	27	12
Move to store, put in store (man-min/t DM)	25	48	76
Total	44	75	88
Straw in trailers			
Capacity of trailer (m³)	18	8	–
Weight of straw in trailer (kg)	850	590	–
Content of DM in straw	87	65	–
DM in trailer (kg)	738	382	–

* Three-man team
[†] Squeeze loader 4 × 2, three-man team working with hay (adapted from data of P. L. Redman)

Provided forage-harvesting occurs when there is moisture (35–50%) in the straw, it can be stored in a silo at a dry-matter (DM) density similar to that of a conventional rectangular bale (100 kg DM per m³). Bulk density of the fresh weight, however, is higher (200–250 kg/m³).

Maize crop residues

The residue from the harvest of maize grain comprises the stem, leaves and usually the husks of the plant. Often, these relatively low-quality residues are grazed *in situ* by beef cattle, but it is quite feasible to harvest them by forage harvester and to ensile the material in much the same manner as described above for straw. Alternatively, if the crop is allowed to dry-out completely in the field, it may be baled.

Sugar beet tops and fodder beet tops

Machinery is now available for the simultaneous harvest of both beet and tops. This greatly reduces the problem of contamination of beet tops with soil, which often occurred when the tops were deposited on the land in a swath at the time of harvesting the roots.

But having overcome the problem of contamination with soil, the direct harvest of beet tops (and crowns) means that a relatively wet material (14–20% DM) is ensiled, which may not ferment well and which will certainly give rise to large quantities of effluent. The former problem may be overcome by the use of an effective additive and the latter by ensiling the beet tops in mixture with, or on top of, straw. A trial in Denmark showed that, by mixing beet tops with 15 per cent chopped straw by weight, the DM content of the mixture was increased to a level at which no effluent loss occurred. But the addition of straw was reflected in a marked reduction in the digestibility of the organic matter of the mixture, compared with beet tops ensiled alone. This reduction may, however, be partially alleviated by using alkali-treated straw.

Brewers' grains and distillery wastes

Wet brewers' grains and wet distillery wastes (e.g. draff) are popular feeds of relatively high nutritive value. The grains are normally conserved by ensilage, and preserve well in clamp and bunker silos. The guidelines described in Chapter 3 for grass crops apply equally to these by-products, especially the need to seal the silo completely to prevent moulding. The composition of ensiled brewers' grains is given in Table 7.2.

The apparently high DM content of ensiled brewers; grains reflects the considerable loss of liquid that occurs during storage. Over 20 per cent of the weight of the material can be lost during storage, but this is mainly water, which is external to the tissues of the grains. To reduce the loss of liquid, a layer of dry straw may be placed at the bottom of the silo; alternatively, the grains may be ensiled on top of wilted grass or maize.

Table 7.2 Composition of ensiled brewers; grains

DM (%)*	28
Ash (% of DM)	4.3
Crude protein (% of DM)	20.4
Metabolizable energy (ME) (MJ/kg DM)	10.0

* 20% DM before ensiling
Source: from MAFF *et al.* (1975) *Technical Bulletin 33*, HMSO, London

Dried brewers' grains and dried distillery wastes can constitute a valuable source of undegraded feed protein for cattle, as a result of denaturation of the protein fraction during the drying process.

Sugar beet pulp

Table 7.3   Composition of ensiled or dried molassed sugar beet pulp

	Ensiled	Dried
DM (%)	22.1	90.0
Ash (% of DM)	10.1	8.2
Crude protein (% of DM)	13.2	13.1
Digestibility (D)-value (%)	82.0	–
ME (MJ/kg DM)	12.9	12.5

Source: from Barber W. P. and Lonsdale C. R. (1980) In *BSAP Occasional Publication, No. 3*

The composition of ensiled and dried molassed sugar beet pulp is given in Table 7.3. The products are similar in nutritive value to barley grain, but the nature of the energy is quite different – predominantly digestible fibre rather than starch.

An interesting approach to the conservation of direct-cut grass, developed in France, is to add dried sugar beet pulp to wet grass crops at the time of ensiling, thus increasing the overall content of DM, reducing losses in effluent and, at the same time, increasing the fermentable carbohydrate content of the ensiled mixture. The cost of adding dried beet pulp, whilst considerably higher than if straw were to be used, would be offset by an improvement in fermentation quality and in the overall energy value of the conserved mixture.

Poultry manure

Dehydration has been the commonest method of conservation for caged-layer excreta and broiler litter. This process, carried out at high temperature, kills potential pathogenic organisms (such as *Salmonella*). However, the process is relatively expensive and interest has recently developed in ensiling these by-products with the addition of formalin or formic acid to aid sterilization. With such additives the materials can be preserved safely for use in mixtures with other feeds for cattle.

Cattle manure

Cattle manure is by far the most abundant by-product, and its sheer quantity presents problems of effective disposal or use on many larger livestock farms. The development of systems of recycling cattle manure, pioneered in north America, has not been widely adopted because of the relatively high investment in equipment that has often been necessary. However, mixtures of cattle manure and forage have been successfully preserved by ensiling ('wastelage').

Wastes from abattoirs

Blood and bones from slaughtered animals, and the offals from poultry, are usually conserved by dehydration. They have a high nutritive value (see Fig. 7.1), and are commonly used in the feed compounding industry in mixtures with cereals and cereal by-products.

More difficult to conserve are the digesta from cattle and sheep, because of the need to transport the material, which is only semi-solid in nature. A possibility might be to extrude a proportion of the moisture before attempting to ensile the material.

Upgrading by-products during conservation

A major limitation to the use of conserved by-products as feeds for beef cattle is that many do not contain both energy and protein in adequate concentrations for the growth of beef cattle (see Fig. 7.1). Not surprisingly, a considerable research effort has been put into devising ways in which by-products may be upgraded during, or immediately after, their conservation, so that they may be exploited to a greater extent as feeds for productive ruminants.

Principally, attention has concentrated on straw, and its treatment with alkalis such as sodium hydroxide or ammonia. The object of the exercise is to increase the digestibility and voluntary intake of the treated material. The alkalis degrade a proportion of the cell wall fraction. They also break the ester cross-linkages between lignin and cellulose in the cell wall, thus rendering the cellulose more available for digestion. A further consequence of this chemical activity is a swelling of the plant tissues so that, when the material enters the rumen, the bacterial enzymes gain more rapid access to the cell wall, thus increasing the rate, as well as the extent, of digestion.

Treatment with sodium hydroxide

The effect of treating straw and cattle manure with NaOH prior to ensiling on composition is shown in Table 7.4.

Treatment with alkali was reflected in very large increases in digestibility in both cases, reflecting principally a solubilization of cell wall hemicaellulose, and to a lesser extent a reduction in the contents of cellulose and lignin in the conserved products.

It is important that there is sufficient moisture in straw to allow adequate distribution of the NaOH through the material, and also to dissipate the heat of solution and/or heat of reaction of the NaOH. If the straw is too dry (more than 85% DM), the heat of reaction between NaOH (in concentrated (30% w/v) solution) and straw may be so great that combustion may result. On the other hand, if the material is too wet (less than 30% DM), the alkali will be diluted to the extent that it has little effect on the straw.

NaOH is the most cost-effective chemical for the upgrading of by-products. But three important problems are associated with its use. First, the chemical is difficult to handle with safety on farms. Great care needs to be paid to protecting operators from accidental contact with the alkali, which can cause rapid and serious burns. Particular care must be taken, therefore, to protect the eyes.

Table 7.4   *Treatment of barley straw and cattle manure* with sodium hydroxide: composition and disgestibility in vitro of the ensiled products*

	Barley straw				Cattle manure	
Level of NaOH (% of DM)	0	2.5	5.0	7.5	0	7.5
Dry matter (%)	56	58	58	59	16	17
pH value	5.7	7.2	9.0	10.1	7.6	7.0
Organic acids (% of DM)	5.1	4.0	2.5	3.2	1.8	15.0
Total nitrogen (% of DM)	0.46	0.45	0.45	0.44	2.1	1.9
Ammonia-N (% of total N)	6.3	3.3	1.3	2.1	9.5	12.8
D-value (*in vitro*) (%)	41	52	62	63	14	53

* Manure = faeces + urine + straw bedding

Figure 7.4
Treating straw with sodium hydroxide. The straw is turned in the mixer-wagon as a solution of sodium hydroxide is pumped at low pressure through holes in an alkathene-pipe frame mounted on top of the wagon

Secondly, addition of NaOH to by-products involves a large increase in the sodium content of the material. Cattle have to excrete the additional sodium via the urine and output of urine is increased to remove the load of sodium. In extreme cases it is possible to detect damage to the kidneys associated with the increased sodium in the blood. The practical consequence is an increased requirement for bedding if the cattle are not housed on slatted floors.

Thirdly, the increase in digestibility associated with the addition of NaOH to straw exacerbates the problem of its low content of N, since the demand for N by the rumen microbial population is increased, along with the potential digestibility of the feed. Thus, in addition to NaOH, there is need to supply additional N in the form of either protein or non-protein N, to maintain the necessary balance between available energy and available (degradable) N in the rumen (Fig. 7.2).

Treatment with ammonia

Since NH_3 is an alkali, which also contains N, it is an obvious choice for use with by-products that are low in N. It is commonly used as a fertilizer in many countries and is therefore readily available on farms.

In Scandinavia, a process for treating straw with 35 kilograms NH_3 per tonne DM has been developed for use by contractors on farms. Baled or loose straw is stacked on a bottom sheet of polyethylene and enclosed in a top sheet sealed to it. A perforated pipe is thrust into the centre of the stack after filling, to allow the injection of NH_3 into the stack from a bulk tanker (see Fig. 7.5 and 7.6). After 2 months of storage the stack is opened for use.

About one-third of the NH_3 is trapped by the straw, raising the crude protein content (N × 6.25) from 3 to 4 per cent of the DM to 10 to 11, equivalent to that of medium-quality hay. Digestibility is improved to almost the same extent as with NaOH (about 15 percentage units on average), but the maximum effect is not

Figure 7.5   A stack of straw for injection with aqueous ammonia. Before injection the stack must be effectively sealed so that no ammonia escapes. The stack should hold about 30 tonnes, just less than 2000 bales.

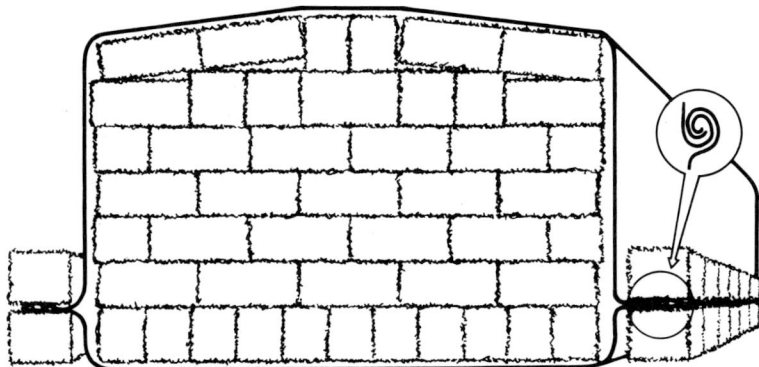

Figure 7.6   Injecting anhydrous ammonia into large round bales of straw as they are conserved in a plastic 'sausage'. The injection point is through the prongs of the fore-end loader.

obtained until after about 2 months of storage.

It appears that with dry straws (10% moisture or less), the response in digestibility to treatment with aqueous NH_3 can be greater than with anhydrous NH_3, at least under laboratory conditions.

An alternative method to the treatment of straw with NH_3 in a sealed stack at ambient temperature is to put bales into an 'oven' through which NH_3 is passed through heated straw. After a period of time to allow the reaction take place, excess NH_3 is removed. The whole process take 24 hours and bales can be treated the day before they are given to the cattle.

Treatment with urea

Urea may be used as a safer alternative to NH_3, although successful upgrading depends upon its conversion to NH_3 during storage. As with NH_3, there is likely to be a substantial increase in the N content of the treated product.

Figure 7.7   Machine for injecting solutions of urea or other liquid feed additives into straw. Packs of 8 or 10 bales are loaded by squeeze-loader and the solution is injected into four bales at a time through 112 vertical probes

Solutions of urea, essential minerals and vitamins are available to add to straw, although here the object is to rectify the low content of N and the low mineral content of the material, rather than to increase digestibility. The urea solution may be added by watering-can, or injected using a purpose-built machine.

Performance of beef cattle given conserved by-products
Straw

There is undoubtedly scope for straw to comprise a major source of feed energy for those types of beef cattle that are not required to gain in weight at a rapid rate. For example, pregnant beef cows have been given diets comprising straw, limited grain, and a supplement of urea and minerals (as a liquid additive mixed into the straw before feeding). (see Table 7.5). The cows consumed adequate quantities of straw and calved normally. Intake of ME was increased by 8 per cent by either treatment of straw with NaOH or by addition of the liquid urea/mineral supplement, and by 18 per cent following both alkali-treatment and addition of the urea/mineral supplement.

Straw can also be used effectively in the diet of intensively reared beef cattle, particularly if the straw has been upgraded during conservation. The feeding value of such straws has been studied at the Rowett Research Institute, Aberdeen. The data in

Table 7.5   *Intake and digestibility of straw diets by pregnant Hereford crossbred beef cows**

	Untreated straw	NaOH-treated straw[†]	Untreated straw + urea/ mineral supple- ment	NaOH-treated straw + urea/ mineral supple- ment
Intake of DM (kg/day)				
Straw	4.51	4.72	5.01	5.08
Barley grain	1.67	1.67	1.67	1.67
Digestibility of organic matter (%)				
Whole diet	56	60	57	62
Straw	46	50	47	53
Intake of ME (MJ/day)	53.5	58.0	58.0	63.2

* 472 kg live weight, initial condition score 2.3
[†] 2.5% of straw DM
[‡] 4% of straw DM
Source: from Bass, Jean M. *et al.*, (1980) *Animal Production* **30**:13

Table 7.6, from Dr J.F.D. Greanhalgh and his co-workers, show the levels of performance achieved by finishing beef cattle given a complete diet of straw and concentrates. The levels of live-weight gain were slightly higher for cattle given straw that had been ensiled after treatment with NaOH, rather than treated immediately before being given to the cattle.

Table 7.6   *Performance of finishing beef cattle given a complete diet of barley straw (40%) and concentrates (60% of total dry matter)*

	Untreated Straw	Treated with NaOH*	
		Feeding	Before ensiling
Intake of DM (kg/day)	8.60	9.7	10.2
Live-weight gain (kg/day)	0.78	1.03	1.08

* 8% of straw DM

Treated straw can also be used to replace grass silage or hay in the diet of growing beef cattle and this tactic may be particularly attractive in dry years, when yields of grass are low and winter supplies of conserved forage are limited. In a trial in Norway, young bulls gave quite reasonable rates of gain when given NH_3 treated straw instead of grass silage in the diet (Table 7.7); differences in gain largely reflected differences in estimated energy intake.

Table 7.7   *Replacement of grass silage by ammonia-treated straw in the diet of growing bulls*

	Grass Silage	Ammonia-treated straw
Initial live weight (kg)	284	284
Final live weight (kg)	451	436
Intake of DM (kg/day)		
Grass silage	4.3	–
NH$_3$-treated straw	–	3.8
Concentrates	2.0	2.0
Hay	1.0	1.0
Relative intake of energy	100	80
Live-weight gain (kg/day)	1.02	0.84

Source: from Arnason, J. and Mo, M. (1977) *3rd MAFF Straw Conference*

However, even with upgraded straw, there is a limit to the extent to which straw can replace other conserved feeds (such as silage) in the diet. Thus the weight gain of calves, given treated straw in different proportions with grass silage, decreased as the proportion of straw in the diet was increased (Fig. 7.8). It is interesting to note that in this trial the calves performed particularly well on the control diet of grass silage supplemented with soya bean meal and that one-third of the silage could be replaced by straw whilst still maintaining a reasonable rate of live-weight gain.

Maize crop residues

The residues that remain following the harvest of the maize grain crop can be a useful source of winter feed for beef cattle. Results

Figure 7.8   Replacement of grass silage by sodium hydroxide-treated straw. The effect on live-weight gain of young Friesian calves

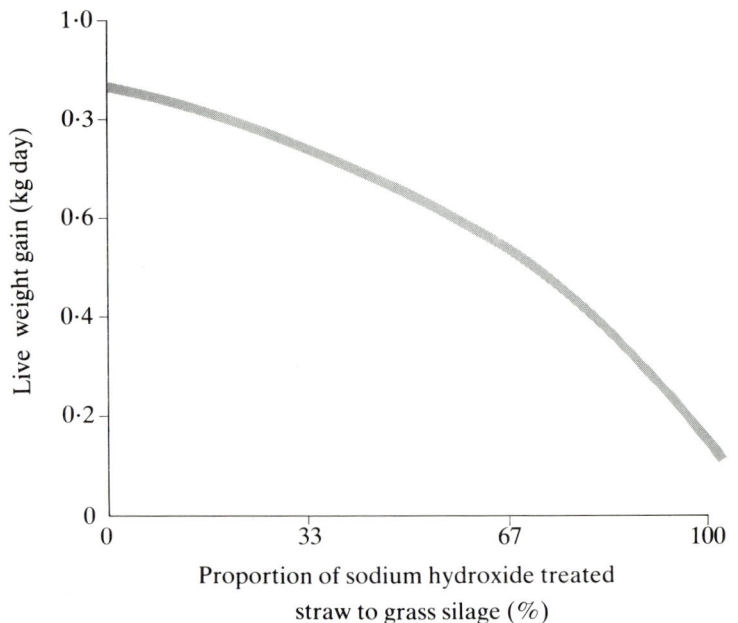

Live weight gain (kg day)

1·0

0·3

0·6

0·4

0·2

0

0   33   67   100

Proportion of sodium hydroxide treated
straw to grass silage (%)

from trials in the USA show that quite acceptable levels of feed intake and live-weight gain can be achieved by growing beef cattle given maize crop residues, provided they are upgraded by treatment with alkali. Performance was similar for cattle given maize silage, alkali-treated ensiled husks (husklage) and alkali-treated maize cobs (Table 7.8).

Sugar beet pulp and citrus pulp

The relatively high energy value of sugar beet pulp and citrus pulp (Fig. 7.1; Table 7.3) indicates that they should support high rates of growth in cattle. Trials in Belgium by Dr C. Boucqué with growing Red and White Belgian bulls have confirmed that pellets of dried sugar beet pulp, given as the sole source of energy with 0.85 per cent urea (to raise the crude protein from 10.8 to 14.0% of the ration DM) and 4 per cent of a mineral/vitamin supplement, can give levels of growth as high as 1.3 kilograms live-weight gain per day. In another trial with younger bulls of the same breed, a diet of 50 per cent dried citrus pulp and 50 of a barley, soya bean concentrate (containing meal, molasses

Table 7.8   Performance of beef cattle (200 kg initial live weight) given maize crop residues

	Maize silage*	Maize husklage†	Maize cobs	
			Untreated	Treated‡
Intake of DM (kg/day)	7.3	7.2	4.1	5.4
Live-weight gain (kg/day)	0.75	0.75	0.30	0.73

* 90% of ration
† Treated with 3% NaOH, 1% calcium hydroxide; on a DM basis. Husklage comprised 80% of ration
‡ Treated with 4% NaOH; on a DM basis. Cobs comprised 80% of ration
Source from Klopfenstein, T. (1978) *Journal of Animal Science* **46**:841

minerals and vitamins) gave growth rates very slightly lower, (1.2 kg live-weight gain per day).

When dried beet pulp and citrus pulp can be obtained at a cost that is competitive with barley grain, they can clearly form the basis of an intensive system of production.

Manures The performance of beef cattle given diets comprising mixtures of forage, by-products or feed grain, and excreta has been examined. In the case of poultry excreta, live-weight gains tends to be lower with diets that contain the excreta than with diets that do not. The average reduction in live-weight gain was 4 percentage units per 10 per cent inclusion of excreta in the diet DM, although there was considerable variation in the response to inclusion of excreta in the diet (Fig. 7.9). In one experiment, in which the control diet was deficient in nitrogen, performance was improved substantially by the addition of excreta.

Beef cows given an ensiled mixture of feedlot cattle manure and maize stover gained weight at almost 1 kg/day during a 56-day period of feeding prior to calving. Calving performance was similar to cows given ensiled maize crop residues supplemented with maize grain and a protein balancer. The excreta silage was palatable and consumed in quantities that were more than adequate to maintain the required nutrient intake of cows in late pregnancy.

As with poultry manure, the majority of trials in which diets containing cattle manure were compared with similar diets

Figure 7.9   Live-weight gains by beef cattle given diets with or without inclusion of poultry excreta. Points on the y=x line indicate no change associated with inclusion of manure in the diet; points below the line indicate a reduction in performance

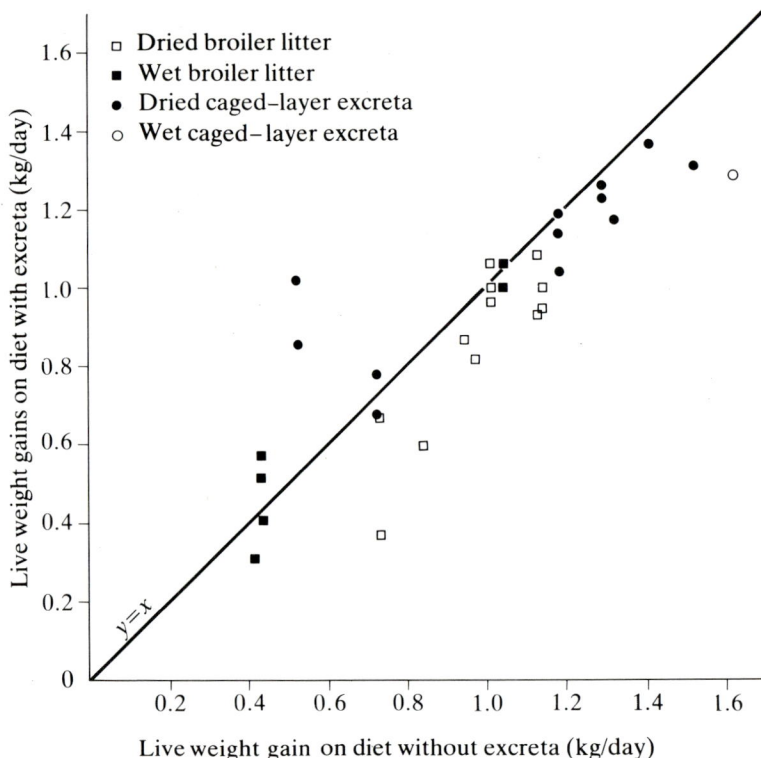

without manure showed a descrease in beef cattle growth; the average decrease in gain in the trials summarized in Fig. 7.10 was 28 percentage units per 10 per cent inclusion of excreta DM in the diet DM (up to 30%).

It appears that manure from cattle given high-concentrate diets can be recycled to cattle and that ensiling is a suitable method of conservation prior to feeding. However, much less is known about the possibilities for recycling the manure from cattle given forage diets, which is likely to be of very low value. A further problem is that, in a recycling system involving manures, indigestible plant cell wall material, particularly lignin, accumulates. This means that, effectively, manure can only be recycled once;

Figure 7.10   Live-weight gains by beef cattle given diets with or without inclusion of cattle excreta. Points on the $y=x$ line indicate no difference associated with inclusion of manure in the diet. Points below the line indicate that performance was reduced

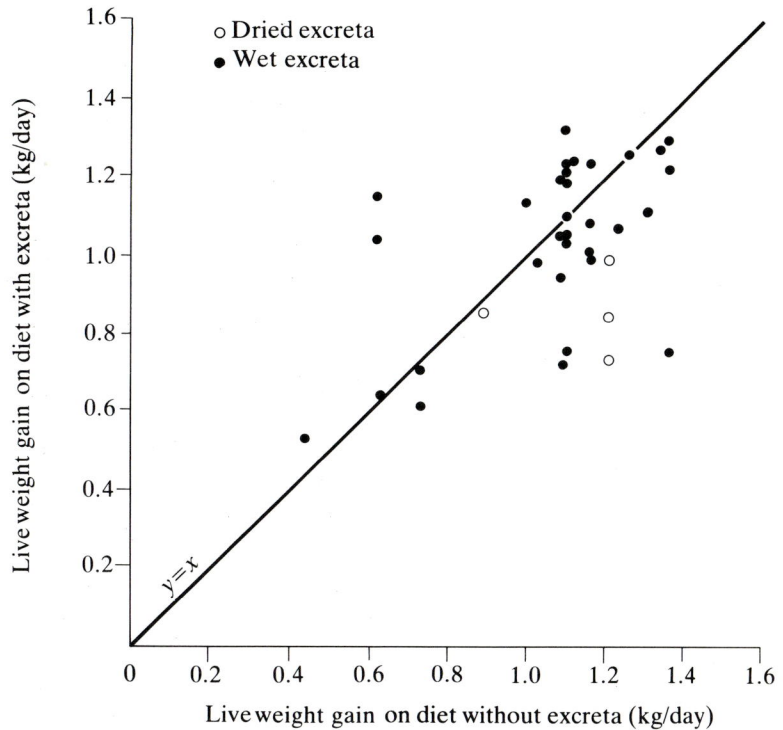

the manure from animals given diets containing excreta then has to be disposed as fertilizer on the land.

There may be possibilities for treating manure with alkali before it is recycled; this may reduce the extent of accumulation of indigestible fractions, but research is needed to demonstrate the efficacy of such treatment.

Chapter 8 New opportunities

Complementary beef production

The foregoing chapters have illustrated the central theme of this book: that beef cattle can use conserved forages and by-products in a productive way to produce food for the human population. In so doing they do not compete for land or feed with humans, rather they co-exist (as indeed they did in the early days of domestication) in a complementary manner, living on land and consuming feed that the human population cannot use.

The long-term prognosis is that, as the demand for human food increases, the increased pressure on grain to supply the human population will lead to shortages in the supply of grains for use as animal feed. These shortages will occur sporadically when, as a result of adverse weather, major grain-consuming nations enter the world market to purchase grain to meet the shortfall in their home-produced supply. When such events occur, the sectors of the live stock industry that will be greatest affected will, of course, be those that use the most grain (pigs and poultry). But since temporary shortages in supply are reflected in increases in the cost of grain to all livestock enterprises, beef producers will also suffer.

It therefore seems prudent to plan systems of beef production so that they are not only biologically efficient (because of the association between biological and economic efficiency) but also less reliant on cereal concentrates than before.

This argument can be taken a stage further. Beef systems can be assessed in terms of their efficiency of use of cereal concentrates. Clearly, the most efficient are those that use none at all. But if the criterion of efficiency is the amount of live weight or carcass gain per unit of concentrate consumed, if weight gain is serverely reduced at low levels of concentrate input then efficiency may be no better than when large quantities of concentrates are eaten and rapid gains are achieved.

Increasing output of beef per unit of concentrates

The concept of defining efficiency as output of weight gain per unit of concentrate lays emphasis on sustaining rapid animal growth from grazed and conserved grass, and forage crops,

Table 8.1   Current levels of output of carcass beef per unit of concentrate input

System	Carcass beef per 100 kg concentrate dry matter (DM) (kg)
Cereal beef	15
Grass/cereal beef	27
Grass beef	24

appropriately, but not excessively, supplemented with concentrates. The different systems of beef production are compared in terms of carcass weight per unit of concentrate in Table 8.1.

Surprisingly, the so-called grass-based systems, in which concentrates (mainly barley) are used to supplement conserved forages, are not markedly more efficient than cereal beef, since cattle are given about 1 tonne of concentrates per head in both the grass/cereal and grass beef systems.

Some examples of the scope for increased efficiency are shown in Table 8.2. They are from work in Europe with either rapidly-growing bulls given maize silage or dried sugar beet pulp, or steers given high-quality grass. It is interesting to note that these efficiencies are similar to the efficiency for broiler chicken production, which is 50 kilograms carcass gain per 100 kilograms concentrate feed. In this case, very rapid growth, mainly of lean tissue, is achieved from a concentrate diet. The beef animal does not reach a similar level of efficiency until a substantial proportion of the total diet dry matter (DM) (about 75%) is from forage.

Table 8.2   Intensive beef production with low levels of concentrate input

System	Live-weight gain (LWG) (kg/day)	Concentrate DM (kg per head)	Carcass beef per 100 kg concentrate DM (kg)
Maize silage	0.9	495	51
Grass	0.8	530	48
Dried sugar beet pulp	1.2	620	46

It would appear quite feasible to use grass silage to produce finished cattle at an age and weight similar to that achieved in the maize silage beef system, and with a similar low input of concentrates, (see Ch. 6). This means a programmed management of grass conservation from land that is not grazed. The target digestibility-value at cutting should be at least 68 per cent and every effort must be made to preserve the crop with minimal nutrient loss. Since the system relies heavily on the production of conserved grass of consistently high quality in adequate quantity, it is appropriate to consider some possible opportunities to improve both the nutritional quality of conserved forages and the efficiency of the conservation process itself, especially with regard to the use of energy.

Reducing the use of support energy in forage conservation

'Support energy' (i.e. energy from non-renewable sources, such as fossil fuels) costs money. Energy-accounting is really only another way of analysing the financial costs of a particular process. But since energy is a scarce and costly resource, in the same way as are land, labour and capital, it is relevant to identify the principle inputs of support energy and to consider possible opportunities for making reductions in its use in the conservation of forages.

The principle inputs of support energy into forage conservation are given in Table 8.3. It is clear that the major inputs of support energy to grass are fertilizer nitrogen (which carries a support-energy cost of 80 MJ/kg N), machinery for field operations and fuel for artificial dehydration. Energy for field operations is higher for ensiling than for haymaking, despite the fact that the production of field-dried hay may involve repeated tedding and turning of the crop. The difference is due to the higher energy cost of the larger tractors and forage harvesters used in silage-making.

Although losses during haymaking are generally greater than those incurred in the production of silage (Ch. 3), haymaking gives a higher ratio of metabolizable energy (ME) output to support-energy input than ensilage, but the output of ME is low.

High-temperature dehydration, by contrast, consumes a very large amount of energy as fuel to dry the crop. So much energy is used in this way that it exceeds the output of ME, despite low losses during the conservation process.

Table 8.3   Support energy use in the conservation of grass

	Artificial dehydration (wilted grass, oil-fired burner)	Hay (field-dried)	Silage (direct-cut + additive)
Number of cuts	5	2	3
Output of metabolizable energy (ME) (GJ/ha)	112	44	80
Input of support energy (GJ/ha)			
Fertilizer*	38	10	19
Field operations†	12	5	13
Capital depreciation	11	–	–
Fuel/additive for silage	82	–	7
Storage/auxiliary equipment	10	1	2
Total	153	16	41
Output: input	0.73	2.75	1.95

* 400, 100 and 200 kg nitrogen per ha for artificial dehydration, hay and silage, respectively, + 100 kg P_2O_5 and 150 kg K_2O per ha
† Including depreciation of machinery

Some alternative methods of conservation are given in Table 8.4, to illustrate the scope for energy-saving in forage conservation.

The use of slurry instead of inorganic fertilizers radically alters the ratio of the output of ME to input of support energy so that, in the case of hay made from crops that have received no fertilizer the 'energy ratio' is as high as 5 to 1, although output is lower than for conventional hay (by 20%) and only about half that of silage.

The substitution of oil by straw as fuel for artificial dehydration (also using slurry) raises the energy ratio so that it approaches that achieved in conventional silage-making, but such an approach has yet to be adopted on a commercial scale.

The production of maize silage by using slurry (40 t/ha) is particularly attractive. Not only is the energy ratio high (Table 8.4), but the output of ME per hectare (70 GJ) and the concentration of ME in the conserved product (10.7 MJ/kg DM) are also relatively high.

Table 8.4 Some possible methods of forage conservation with low use of support energy

Crops	Artificial dehydration	Hay	Silage	
	Grass Grass/clover Lucerne Maize	Grass Grass/clover Lucerne	Grass Grass/clover Lucerne	Maize
Fertilizer	Slurry	Slurry	Slurry	Slurry
Fuel	Straw	–	–	–
Number of cuts	4	2	3	1
Output of ME (GJ/ha)	75	35	60	70
Input of support energy (GJ/ha)				
Field operations	12	6	15	13
Capital depreciation	10	–	–	–
Storage/auxiliary equipment	9	1	2	2
Storage of straw	15	–	–	–
Total	46	7	17	15
Output:input	1.63	5.00	3.53	4.67

Another possibility is to produce barn-dried hay by solar-heated air, rather than by electrically driven fans. This approach has the attraction of both reducing the support energy cost of conventional barn-dried hay and also reducing field losses. It is likely that the installation of equipment for the solar-drying of hay will be competitive with conventional barn-drying equipment.

Legumes   Much research effort has been expended on understanding the factors that influence the relative feed value of grasses and legumes; we now know that legumes are generally eaten in greater quantities than are grasses of similar digestibility. There is a considerable amount of evidence to show positive responses in the live-weight gain of beef cattle to legumes or grass/legume mixtures, compared with that achieved with similar cattle given grass alone (Fig. 8.1).

A recent comprehensive trial, completed by A. T. Stewart at Greenmount and Loughry in Northern Ireland, involved a

Figure 8.1 Live-weight gain (kg/day) by beef cattle given legume or legume plus grass forages compared with grass alone. Almost all the points lie above the line of equal response ($y=x$), showing superiorty to the legume over a wide range of live-weight gain. From Thomson, D. J. (1978) In *British Grassland Society Occasional Symposium, No. 10*

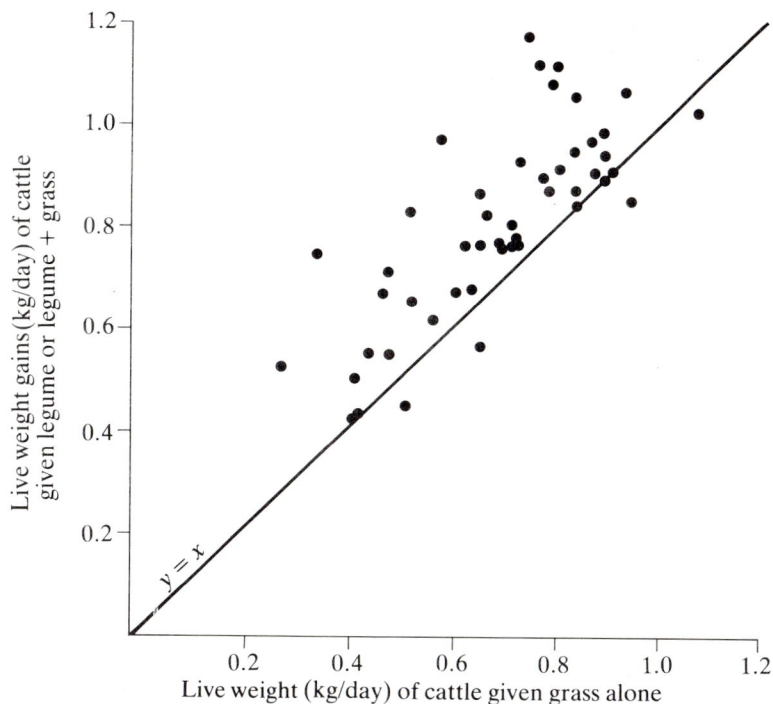

comparison of grass (300 kg N per ha) with grass and white clover (50 kg N per ha) for beef cattle in an 18-month grass/cereal system of production. The main results are given in Table 8.5.

Cattle were stocked at 3.33 animals per hectare on the low-N grass/clover sward and at 4.53 per hectare on the high-N grass sward. Silage made from areas surplus to conservation was given with 1.5 kilograms cereal grain supplement per head per day during the winter. Performance was 70 g/day higher for cattle on the low-N sward, but this difference only occurred at pasture. Output of carcass weight per hectare was 16 per cent lower for the low-N treatment but, because of the reduced fertilizer costs, the average gross margin per hectare was similar to that obtained from the high-N grass system. The working capital required per

Table 8.5 *Output of beef, gross margin and efficiency of use of support energy in 18-month grass/cereal beef production from grass/clover swards compared with grass swards*

	Grass	Grass/clover	Difference
LWG (kg/day)*			
6–12 months of age, at pasture	0.84	0.91	+0.07
6 months to slaughter	0.77	0.84	+0.07
Carcass weight (kg/ha)	596	500	−96
Working capital (£ per ha)	1 040	720	−320
Gross margin (£ per ha)	514	518	+4
Gross margin (% of working capital†)	50	72	+22
Energy ratio (output : input)	0.15	0.23	+0.08

* Mean of 4 years, 1977–80
† Gross margin % of working capital for grass/cereal 18-month beef production averaged 57% on recorded farms in the UK, 1972–77

hectare to purchase cattle and fertilizer was 31 per cent lower for the low-N system and, in consequence, gross margin as a percentage of working capital invested was substantially higher. Efficiency of use of support energy was over 50 per cent greater for the grass/clover than for the grass system.

The lower output of carcass weight per hectare reflects the lower yield of the grass/legume crop, compared with the high-N grass sward. A similar situation applies to legumes grown specifically for conservation; the yield of current varieties of lucerne and red clover does not match that that is achievable from well fertilized ryegrass. This problem is exacerbated by the fact that, at similar digestibility, consumption of DM from legume silage may be as much as 20 per cent higher than that of grass silage, at least when the silages are offered to young cattle as the sole feed (Table 8.6). As a result, the level of live-weight gain by calves given red clover silage is substantially higher than that achieved by comparable animals given grass (Fig. 8.3). This, in turn, partially redresses the balance, since savings can be made with cattle given legume silages in the cost of concentrate supplements.

It is interesting to note that, despite claims that the ME of legumes may be utilized more efficiently than that of grasses, there was no difference between the two species in the relationship between digestible organic matter (DOM) intake and live-

Figure 8.2   To achieve improved beef cattle performance from grass/clover swards, the proportion of clover should exceed 25 per cent of the total crop dry matter

Table 8.6   Intake and performance of young Friesian beef cattle given silage made from red clover or ryegrass

	No supplement		With supplement*	
	Ryegrass	Red clover	Ryegrass	Red clover
Mean live weight (LW) (kg)	132	156	172	184
D-value of silage (%)	62	63	73	72
Intake of DM (g/kg LW per day)	22.4	27.1	26.3	28.9

* Barley, given at 11.5 g DM per kg LW per day
† Mean of two cuts for each species
Source: after Thomas, C. *et al.* (1981) *Animal Production* **32**:149

weight gain in the trial reported in Fig. 8.3. Thus the higher weight gains of calves given the clover diets were a consequence of higher intakes rather than higher efficiency of utilization of

Figure 8.3   Performance of young beef cattle given silages made from red clover or perennial ryegrass in relation to intake of digested organic matter. After Thomas, C. *et al.* (1981) *Animal Production* **32**: 149

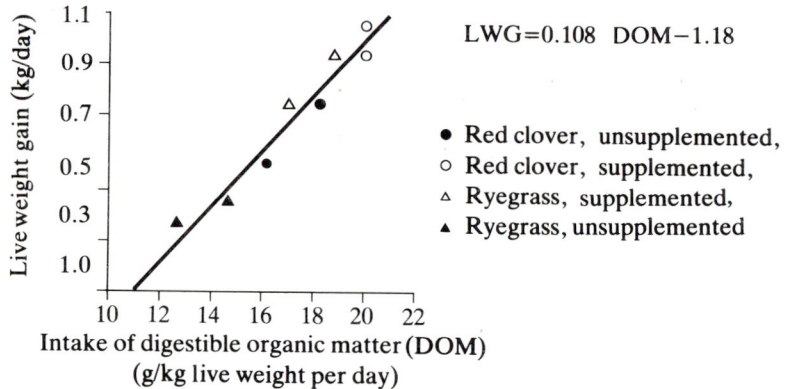

DOM. A further relevant point is that the differences between the two species in intake were much smaller when the silages were supplemented than when they were not. This reflects the higher substitution value of the legume silage.

Thus there appear to be real possibilities of developing energy-efficient and capital-efficient systems of beef production that rely, to a much greater extent than before, on legumes. A challenge to the research worker remains, however, to produce legume crops and grass/legume mixtures that will equal the yield of pure-grass swards. Until this is achieved, the penalty of reduced output per hectare of land from legume-based beef production remains a major limitation, particularly when beef competes with other enterprises for land on the farm.

Legumes are used successfully in combination with maize silage, particularly in the feedlots of north America. Here the two crops are grown not only because they are complementary in relation to their composition as feeds for beef cattle; the legume can also confer yield benefits to subsequent crops because it leaves residual N in the soil.

Whole-crop cereals

The urge to use straw in beef cattle diets is prompted by the belief that it is a cheaper source of ME than grass, even after a process of upgrading. In areas where arable crops predominate, there is the further desire to maximize the area of land devoted to cereal-grain, since gross margins per hectare obtained from grain are usually higher than those from beef.

But straw alone, even after upgrading, is insufficiently high in energy content and is not eaten in great enough quantity to promote rapid growth in beef cattle. Supplementary grain is needed in the diet. This raises the question as to why one should go to the trouble of harvesting the grain separately from the straw, conserving each separately, then recombining the two parts when the crop is given to the animal. The answer is that often the desire is to sell the grain (or a proportion of it) as a cash crop and that some straw is needed for bedding. Also, by separating grain from straw, the ratio of the one to the other can be altered, as required, at the time of diet formulation.

With increasing costs of farm machinery, the cost-effectiveness of the combine harvester is coming under close scrutiny. It is an expensive, inflexible, under-used piece of equipment. Experts in farm mechanization are now seriously doubting the wisdom of investing in such a machine when other approaches may be adopted.

One alternative is to harvest and ensile the cereal crop as a whole, accepting that the policy is to add value to home-grown grain (and straw) by converting it into beef. Harvest is by forage harvester with a direct-cut attachment and the crop is conserved in the same way as conventional grass silage. The optimal stage of maturity for the harvest of whole-crop wheat and barley is when the crop has reached the dough-ripe stage of grain maturity. At this stage the DM content of the whole-crop is between 40 and 50 per cent.

Losses of grain from forage-harvesting the whole crop appear to be similar to those incurred by using a combine harvester. With the latter, a higher loss before harvest, due to grain shedding from the more mature plants, is counterbalanced by a lower loss at harvest. Some damage occurs to the grain as a result of the action of the cutting knives of the forage harvester, but it appears that less than 10 per cent of grain is damaged in this way. Thus it is important in making whole-crop cereal silage that the grain is softened during the ensiling process. This usually happens if the crop is harvested at the dough-ripe stage of grain maturity.

The performance of beef cattle given whole-crop cereal silages

Table 8.7   Performance of Hereford and Aberdeen-Angus steers given whole-crop cereal silages

Silage	Trial 1*	Trial 2*			
	Maize[†]	Barley[‡]	Wheat[‡]	Oat[†]	Sorghum[†]
Number of steers	18	18	36	36	12
Initial LW (kg)	295	291	292	293	210
Final LW (kg)	397	385	374	336	282
Intake of DM (g/kg LW per day)	25.1	26.2	25.9	21.2	27.7
LWG (kg/day)	1.15	1.06	0.91	0.48	0.76

* Silage comprised 84 and 73% of total diet DM in trials 1 and 2, respectively
[†] Spring-sown cultivars
[‡] Winter-sown cultivars
Source: from Bolsen, K. K. (1980) In *British Grassland Society Occasional Symposium, No. 11* and Oltjen, J. W. and Bolsen, K. K. (1980) *Journal of Animal Science* **51**:958

is given in Table 8.7. Live-weight gains reflected the proportion of grain in the silage, which was highest for maize and, in these particular trials, lowest for oats. Intake of DM was also depressed in the case of the oat silage. Gains exceeded 0.9 kg/day for maize, barley and wheat silages, whilst that from forage sorghum was somewhat less. With the exception of the maize crop, which was harvested at the hard-dent stage of grain maturity, all the other whole-crop cereals were harvested at the dough-ripe stage of grain maturity.

If responses in digestibility and performance occur following treatment of straw with alkali (Ch. 7), then there may be benefits from treating whole-crop cereals in similar manner. However, although alkali-treated material is stable when preserved by ensiling, when the silo is opened it can go mouldy rapidly and become unpalatable. Further work is needed to confirm the efficacy of treating whole-crop cereals with alkali on a farm scale.

An alternative to ensiling the whole cereal crop is to harvest the crop as previously described with a forage harvester, but then to separate it into its major component parts by 'winnowing' in the barn (see Fig. 8.4). The crop is elevated to a fan, which blows the crop over a number of hoppers. According to the density and speed of air flow, a number of fractions can be separated.

Grain (fraction 1) may be sold or used for animal feed on the

Figure 8.4   Schematic diagram of a process for separating whole-crop cereals into major components in the barn

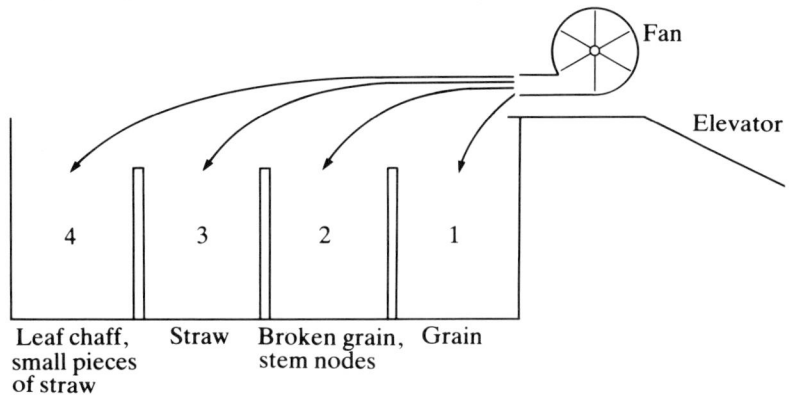

farm, as required. Broken grain (fraction 2) may be used as animal feed. Fraction 3, the straw, may be used for feed, bedding or fuel, as required. Fraction 4 may be recombired with fraction 2 for use as feed. It is worth noting that, with conventional combine harvesting and baling, this fraction is usually lost.

Upgrading grass during conservation

It is commonly lamented that conserved grass does not support the same levels of animal performance as when the same fields are grazed in summer. Such lamentation is unjustified; like is not being compared with like. Under grazing, cattle are usually offered at least twice as much grass at any one time as they are likely to eat. They reject the less palatable fractions, which tend to be of lower digestibility than those that are selected and eaten. By contrast, when cattle are offered silage or hay they are expected to eat it all. Further, the losses that occur during conservation comprise the most-digestible cell contents, rather than the less-digestible cell wall fraction. In addition, crops are commonly harvested for conservation, particularly those for hay, at a more advanced stage of growth than those that are grazed. Thus it is hardly surprising that large quantities of cereal-grain are used in the winter period to maintain the growth of beef cattle and to produce a 'finished' animal.

Fortunately, the time that elapses between harvesting the crop for conservation and giving the product to the animal introduces opportunities for making a net improvement in nutritive value, so that the conserved product is of superior quality to that of the crop at the time of harvest. In other words, the storage period is used to 'pre-digest' the crop before it reaches the animal.

To date, research in the area of pre-digestion has concentrated on applying to grass the technology developed for straw. The response to alkali is greater with more mature crops than with younger ones of higher initial digestibility (Fig. 8.5). Thus the digestibility of treated hay is similar to that of untreated material made from grass cut several weeks earlier.

It follows that, by upgrading conserved crops, harvest may be delayed to allow yield to accumulate without the usual loss in quality. Stocking rate and output of beef per hectare of land can thereby be increased. Alternatively, if hay is the desired conserved crop, harvest may be delayed if the weather is not

Figure 8.5   Treatment of grass hay with alkali gives the greatest response in digestibility with mature, low-quality crops – thus harvest may be delayed to allow yield to accumulate without the usual loss in quality

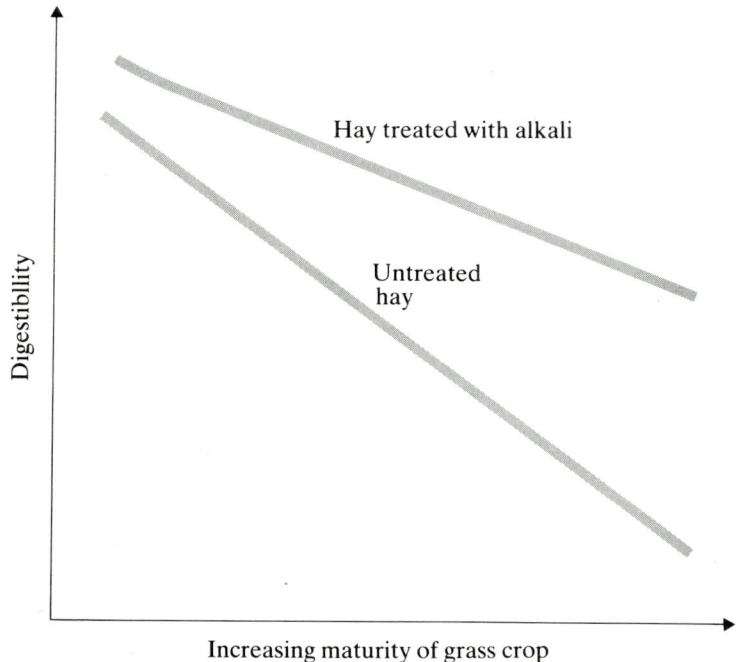

good. The decrease in quality in the more mature, later-cut crop can be recovered during storage.

Future developments will most probably involve the replacement of chemicals that are difficult to handle on farms by others that are not. Urea, for example, may hold promise as a way of producing ammonia *in situ* by hydrolysis, using urease enzymes present in the plant material.

There may be even greater opportunities for pre-digesting crops during conservation by the addition of enzymes that will attack the plant cell wall constituents. With the advent of genetic engineering, it is quite feasible that such enzymes will eventually be 'harvested' from microbial populations in amounts large enough so that they can then be added to crops to increase nutritive value during the conservation process. Such a development could transform the efficiency with which beef is produced from silage and other conserved forages.

Further reading

Allen, D., and Kilkenny, B. (1980) *Planned Beef Production*, Granada Publishing, London.

Lasley, J.F. (1981) *Beef Cattle Production*, Prentice Hall, Englewood Cliffs, NJ.

McDonald, P. (1981) *The Biochemistry of Silage*, Wiley, Chichester.

McDonald, P., Edwards, R.A. and Greenhalgh, J.F.D. (1981) *Animal Nutrition* (3rd edn), Longman, London.

Nash, M.J. (1978) *Crop Conservation and Storage in Cool Temperate Climates*, Pergamon Press, Oxford.

Preston, T.R., and Willis, M.B. (1974) *Intensive Beef Production* (2nd edn), Pergamon Press, Oxford.

Raymond, W.F., Shepperson, G. and Waltham, R.W. (1978) *Forage Conservation and Feeding*, (3rd edn), Farming Press, Ipswich.

Index